Introduction to Probability and Statistics

Introduction to Probability and Statistics

Hubert Murphy

STATES
ACADEMIC PRESS
www.statesacademicpress.com

States Academic Press,
109 South 5th Street,
Brooklyn, NY 11249, USA

Visit us on the World Wide Web at:
www.statesacademicpress.com

ISBN: 978-1-63989-715-5

Cataloging-in-publication Data

Introduction to probability and statistics / Hubert Murphy.
p. cm.
Includes bibliographical references and index.
ISBN 978-1-63989-715-5
1. Probabilities. 2. Statistics. 3. Mathematical statistics. I. Murphy, Hubert.
QA273 .P76 2023
519--dc23

Table of Contents

Preface

I am honored to present to you this unique book which encompasses the most up-to-date data in the field. I was extremely pleased to get this opportunity of editing the work of experts from across the globe. I have also written papers in this field and researched the various aspects revolving around the progress of the discipline. I have tried to unify my knowledge along with that of stalwarts from every corner of the world, to produce a text which not only benefits the readers but also facilitates the growth of the field.

Probability and statistics are two sub-disciplines of mathematics. Probability is related to the chance of occurrence of an event, whereas statistics is involved in handling data utilizing various techniques. Probability deals with numerical explanations of possible occurrence of an event or likelihood of a proposition being true. The probability of an event is a number between 0 and 1, wherein 0 represents impossibility and 1 represents certainty. The higher the probability, it is more likely that the particular event will happen. Statistics deals with gathering, organizing, interpreting, analyzing, and presenting data. It is a common practice to start with a statistical model or statistical population to be researched when applying statistics to a scientific, social, or industrial problem. This book provides comprehensive insights into the fields of probability and statistics. It presents this complex subject in the most comprehensible and easy to understand language. This book will provide comprehensive knowledge to the readers.

Finally, I would like to thank all the contributing authors for their valuable time and contributions. This book would not have been possible without their efforts. I would also like to thank my friends and family for their constant support.

Hubert Murphy

1

Conditional Probability

1.1 Baye's Theorem

Statement

If $A_1, A_2, \ldots\ldots\ldots A_n$ are partition of a set 's' so that there union in s & E be any other event then;

$$P\left(\frac{A_i}{E}\right)=\frac{P\left(\frac{E}{A_i}\right)P A_i}{\sum_{i=1}^{n} P\left(\frac{E}{A_i}\right)P A_i}$$

Proof:

$$S = A_1 \cup A_2 \cup A_3 \cup \ldots\ldots\ldots \cup A_n$$

$$E = E \cap S$$

$$E = E \cap A_1 \cup A_2 \cup A_3 \cup \ldots\ldots\ldots \cup A_n$$

$$= E \cap A_1 \cup E \cap A_2 \cup \ldots\ldots\ldots \cup E \cap A_n$$

$$P\,E = P[E \cap A_1 \cup E \cap A_2 \cup \ldots\ldots\ldots \cup E \cap A_n]$$

Since $A_1, A_2, A_3, \ldots\ldots\ldots A_n$ are mutually disjoint set.

$$P(E) = P(E \cap A_1) + P(E \cap A_2) + \ldots\ldots + P(E \cap A_n)$$

$$P(E) = \sum_{i=1}^{n} P(E \cap A_i) \qquad \ldots(1)$$

By definition of conditional probability:

$$P\left(\frac{A_i}{E}\right)=\frac{P(A_i \cap E)}{P(E)}$$

$$=\frac{P(E\cap A_i)}{P(E)}$$

$$=\frac{P(E\cap A_i)}{\sum_{i=1}^{n}P(E\cap A_i)} \text{ from (1)}$$

$$P\left(\frac{A_i}{E}\right)=\frac{P\left(\frac{E}{A_i}\right)P(A)}{\sum_{i=1}^{n}P(E\cap A_i)}$$

This result is known as theorem of inverse probability of Baye's theorem.

Problems

1. In a college boys & girls are equal in proportion. It was found that 10 out of 100 boys & 25 out of 100 girls were using same company of 2 wheels. If the student using that was selected at random. Let us know what is the probability of being a boy.

Solution:

Given:

Boys & girls are equal in proportion.

10 out of 100 boys & 25 out of 100 girls were using same company of 2 wheels.

Formula to be used:

$$PE = P A \cdot P\left(\frac{E}{A}\right) + P B \cdot P\left(\frac{E}{B}\right)$$

$$P\left(\frac{E}{A}\right)=\frac{10}{100}=0.1;\ P A = 100\% = 0.5$$

$$P\left(\frac{E}{B}\right)=\frac{25}{100}=0.25;\ P A = 100\% = 0.5$$

Let E be an event of choosing a student from same company of two wheelers:

$$PE = P A \cdot P\left(\frac{E}{A}\right) + P B \cdot P\left(\frac{E}{B}\right)$$

$$=0.5\ (0.1)+0.5\ (0.25)$$

$$PE = 0.35$$

Probability of choosing a student from boy:

$$P\left(\frac{A}{E}\right)=\frac{PA \cdot P\left(\frac{E}{A}\right)}{P\ E}$$

$$=\frac{1\ (0.1)}{0.35}$$

$$=0.2857$$

2. In a school 25% of the students failed in first language, 15% of the students failed in second language and 10% of the students failed in both. If a student is selected at random let us find the probability that (i) He failed in first language if he had failed in the second language. (ii) He failed in second language if he had failed in the first language. (iii) He failed in either of the two languages.

Solution:

Let A be set of students failing in the first language and B be the set of students failing in the second language. We have by data,

$$P(A)=\frac{25}{100}=\frac{1}{4},\quad P(B)=\frac{15}{100}=\frac{3}{20},\quad P(A\cap B)=\frac{10}{100}=\frac{1}{10}$$

$$\text{(i) } P(A/B)=\frac{P(A\cap B)}{P(B)}=\frac{1/10}{3/20}=\frac{2}{3}$$

$$\text{(ii) } P(B/A)=\frac{P(A\cap B)}{P(A)}=\frac{1/10}{1/4}=\frac{2}{5}$$

$$\text{(iii) } P(A\cup B)=P(A)+P(B)-P(A\cap B)=\frac{1}{4}+\frac{3}{20}=\frac{1}{10}=\frac{3}{10}$$

3. The probability that a team wins a match is 3/5. If this team play 3 matches in a tournament, let us determine what is the probability that the team (i) Win all the matches (ii) Win atleast one match (iii) Win atmost one match (iv) Lose all the matches.

Solution:

Let W be the event of winning a match by the team:

$$P(W_1)=P(W_2)=P(W_3)=3/5$$

Let L be the event of losing a match by the team:

Therefore,

$$P(L_1)=P(L_2)=P(L_3)=2/5$$

(i) Probability of winning all the matches:

$$=P(W_1)P(W_2)P(W_3)=27/125$$

(ii) Probability of winning at least one match:

= 1- probability of losing all the matches

$$=1-P(L_1)P(L_2)P(L_3)$$

$$=1-(8/25)=17/25$$

(iii) Probability of winning atmost one match:

$$=P(L_1)P(L_2)P(L_3)+P(W_1)P(L_2)P(L_3)+P(L_1)P(W_2)P(L_3)+P(L_1)P(L_2)P(W_3)$$

$$=\frac{8}{125}+3\left[\frac{3}{5},\frac{2}{5},\frac{2}{5}\right]=\frac{44}{125}$$

(iv) Probability of losing all the matches:

$$=P(L_1)P(L_2)P(L_3)=\frac{8}{125}$$

1.1.1 Random Variables

For simplicity, suppose S is a finite set, so that X and h(X) are necessarily finite valued random variables. Then we can compute the expected value E [h(X)] in three different ways:

$$E[h(X)]=\sum_{s\in S}h(X(s))P(\{s\})=\sum_{s\in S}h(X(s))p(\{s\})$$

$$=\sum_{r\in R}h(r)P(X=r)=\sum_{r\in R}h(r)p_x(r)$$

$$=\sum_{t\in R}tP(h(X)=t)=\sum_{t\in R}tp_h(x)(t)$$

Similarly, we have the following expressions when all these probability distributions have probability density functions.

First, suppose that the underlying probability distribution P on the sample space S has a probability density function f, then under regularity conditions, the random variables X and h(X) have probability density functions $f_x(r)$and $f_h(X)$. Then we have:

$$E[h(X)] = \int_{s \in S} h(X(s)) f(s) ds$$

$$= \int_{-\infty}^{\infty} h(r) f_x(r) dr$$

$$= \int_{-\infty}^{\infty} t f_h(x)(t) dt$$

Law of Large Numbers

Given a large number of repeated trials, the average of the results will be approximately equal to the expected value.

Expected Value

The mean value in the long run for many repeated samples, symbolized as E(X):

$$E(X) = \sum x_i p_i$$

Where,

x_i = value of the ith outcome
p_i = probability of the ith outcome

According to this formula, we take each observed X value and multiply it by its respective probability. We then add these products to reach our expected value. It is also known as a weighted average because it takes the probability of each outcome into account and weighs it accordingly. This is in contrast to an unweight average which would not take the probability of each outcome into account and weigh each possibility equally.

Example:

1. A fair six sided die is tossed. We win $2 if the result is a "1", we win $1 if the result is a "6", but otherwise we lose $1.

Solution:

Given:

Six sided die.

Win $2 if the result is "1".

Win $1 if the result is "6".

Otherwise lose \$1.

Formula to be used:

Probability Distribution for X = Amount Won or Lost

X	+\$2	+\$1	-\$1
Probability	1/6	1/6	4/6

Probability Distribution for X = Amount Won or Lost

$$E(X) = \$2(1/6) + \$1(1/6) + (-\$1)(4/6) = \$-1/6 = -\$0.17$$

The interpretation is that if we play many times, the average outcome is losing 17 cents per play. Thus, over time we should expect to lose money.

Standard Deviation of a Discrete Random Variable

For instance, we may "expect" to win \$20 when playing a particular game, but the spread for this might be from losing \$20 to winning \$60. Knowing such information can influence our decision on whether to play.

To calculate the standard deviation, we first must calculate the variance. From the variance, we take the square root and this provides us the standard deviation. Conceptually, the variance of a discrete random variable is the sum of the difference between each value and the mean times the probability of obtaining that value.

Variance for a Discrete Random Variable:

$$\sigma^2 = \sum\left[(x_i - \mu)^2 p_i\right]$$

Standard Deviation for a Discrete Random Variable:

$$\sigma = \sqrt{\sum\left[(x_i - \mu)^2 p_i\right]}$$

Where,

x_i = value of the ith outcome

$\mu = E(X) = \sum x_i p_i$

p_i = probability of the ith outcome

In these expressions, we substitute our result for E(X) into μ because μ is the symbol used to represent the mean of a population.

Computational Formula

Variance for a Discrete Random Variable:

$$\sigma^2 = \left[\sum \left(x_i^2 p_i\right) - \mu^2\right]$$

Standard Deviation for a Discrete Random Variable:

$$\sigma = \sqrt{\left[\sum \left(x_i^2 p_i\right)\right] - \mu^2}$$

Where,

$$x_i = \text{value of the } i^{th} \text{ outcome}$$

$$\mu = E(X) = \sum x_i p_i$$

$$p_i = \text{probability of the } i^{th} \text{ outcome}$$

Notice in the summation part of this equation, we only square each observed X value and not the respective probability. Also note that the μ is outside of the summation.

Functions of Random Variables

There are three main methods to find the distribution of a function of one or more random variables. They are:

- To use the CDF.
- To transform the pdf directly.
- To use moment generating functions.

Method of Distribution Functions

Suppose the random variable Y has a pdf:

$$f_Y(y) = 3y^2, \ 0 < y < 1$$

And zero otherwise. Suppose we want to find the pdf of U = 2Y+3.

The range of U is 3 < U < 5. Now:

$$F_U(u) = P(U \le u) = P(2Y + 3 \le u) = P\left(Y \le \frac{u-3}{2}\right)$$

$$P\left(Y \le \frac{u-3}{2}\right) = \int_0^{\frac{u-3}{2}} f(y)dy = \int_0^{\frac{u-3}{2}} 3y^2\,dy = \left(\frac{u-3}{2}\right)^3$$

So,

$$F_U(u) = \begin{cases} 0 & u < 3 \\ \left(\frac{u-3}{2}\right)^3 & 3 \le u \le 5 \\ 1 & u > 5 \end{cases}$$

And

$$f_U(u) = \frac{dF}{du} = \begin{cases} \frac{3}{8}(u-3)^2 & 3 \le 5 \\ 0 & \text{otherwise} \end{cases}$$

The general method works as follows:

- U is a function of n random variables $Y_1,\ Y_2,\ \ldots\ldots\ldots\ldots, y_n$.
- Find the region $U = u$ in $(y_1, y_2, \ldots\ldots, y_n)$ space.
- Find the region $U \le u$.
- Find $F_U(u) = P(U \le u)$ by integrating the $f(y_1, y_2, \ldots\ldots, y_n)$ density over the region $U \le u$.
- Find the density function $f_U(u)$ by differentiating $f_U(u)$.

The cdf method is useful for dealing with the squares of random variables. Suppose $U = X^2$, then

$$F_U(u) = P(U \le u)$$

$$= P(X^2 \le u)$$

$$= P(-\sqrt{u} \le X \le \sqrt{u})$$

$$= \int_{-\sqrt{u}}^{\sqrt{u}} f(x)dx$$

$$= Fx(\sqrt{u}) - Fx(-\sqrt{u})$$

So if we differentiate both sides with respect to u, we find:

$$f_U(u) = fx(\sqrt{u})\left(\frac{1}{2\sqrt{u}}\right) + fx(-\sqrt{u})\left(\frac{1}{2\sqrt{u}}\right)$$

$$= \frac{1}{2\sqrt{u}}\left[fx(\sqrt{u}) + fx(-\sqrt{u})\right]$$

So, for example if $fx(x) = (x+1)/2 \quad -1 \le x \le 1$ and zero otherwise and $U = X^2$ then,

$$f_U(u) = \frac{1}{2\sqrt{u}}\left[fx(\sqrt{u}) + fx(-\sqrt{u})\right]$$

$$= \frac{1}{2\sqrt{u}}\left[\frac{\sqrt{u}+1}{2} + \frac{-\sqrt{u}+1}{2}\right]$$

$$= \frac{1}{2\sqrt{u}}, \; 0 \le u \le 1$$

Suppose Z ~ N (0, 1) so that,

$$fz\,(z) = \frac{1}{\sqrt{2}}\exp\left(-\frac{z^2}{2}\right), \quad -\infty < z < \infty$$

$$f_U(u) = \frac{1}{2\sqrt{u}}\left[fx(\sqrt{u}) + fx(-\sqrt{u})\right]$$

$$= \frac{1}{2\sqrt{u}}\left[\frac{1}{\sqrt{2\pi}}\exp\left(-\frac{u}{2}\right) + \frac{1}{\sqrt{2\pi}}\exp\left(-\frac{u}{2}\right)\right]$$

$$= \frac{1}{\sqrt{2\pi u}}\exp\left(-\frac{u}{2}\right)$$

Then $U = Z^2$

1.1.2 Discrete and Continuous Distributions

Types of random variables:

- Discrete Random variables.
- Continuous Random variables.

Two Dimensional Discrete Random Variables

If the possible values of (X, Y) are finite, then (X, Y) is called as two dimensional discrete R.V. and it can be represented by (x_i, y_i), i = 1,2,....,m.

In the study of two dimensional discrete R.V's, we have the following 5 important terms.

Joint Probability Function (JPF) (or) Joint Probability Mass Function.

- Joint Probability Distribution.
- Marginal Probability Function of X.
- Marginal Probability Function of Y
- Conditional Probability Function.
- Joint Probability Function of Discrete R.V'S, X and Y.

The function $P(X=x_i, Y=y_j)=P(x_i, y_j)$ is called as joint probability function for discrete random variable X and Y and it is denote by p_{ij}.

Note:

1. $P(X=x_i, Y=y_j)=P[(X=x_i)\cap(Y=y_j)]= p_{ij}$

2. It should satisfy the following conditions:

(i) $p_{ij} \geq \forall$ i, j

(ii) $\Sigma_j \Sigma_i p_{ij} = 1$

Problems

1. If the joint p.d.f. of the R.Vs X and Y is given by:

$$f(x,y)=\begin{cases} 2, 0<x<y<1 \\ 0, \text{otherwise.} \end{cases}$$

Let us see the p.d.f. of the R.V. $U=\frac{X}{Y}$.

Solution:

Given:

$$f(x,y)=\begin{cases} 2, 0<x<y<1 \\ 0, \text{otherwise.} \end{cases}$$

$$U = \frac{X}{Y}; 0 < x < y < 1$$

Formula to be used:

$$J = \frac{\partial(x, y)}{\partial(u, v)}$$

$$g(u) = \int_{-\infty}^{\infty} h(u, v)\, dv$$

$$U = \frac{X}{Y}; 0 < x < y < 1$$

$$\Rightarrow 0 < x/y < 1$$

$$\Rightarrow 0 < u < 1$$

Let $V = Y \Rightarrow X = UV$

Let the p.d.f of the R.V. $U = \frac{X}{Y}$ is $h(x, y)$.

$$h(x, y) \quad f(x, y) \; |J|$$

Where,

$$J = \frac{\partial(x, y)}{\partial(u, v)}$$

$$y = v \; ; \; x = uv$$

$$J = \begin{vmatrix} \frac{\partial x}{\partial u} & \frac{\partial x}{\partial v} \\ \frac{\partial y}{\partial u} & \frac{\partial y}{\partial v} \end{vmatrix} = \begin{vmatrix} v & u \\ 0 & 1 \end{vmatrix} = v$$

$$h(x, y) = f(uv, v) \; |J| = 2v$$

$\therefore$ The Joint p.d.f of u and v is given by:

$$h(u, v) = \begin{cases} 2v \; ; 0 < u < 1, 0 < v < 1 \\ 0 \; ; \quad \text{otherwise} \end{cases}$$

The marginal p.d.f of v is given by:

$$g(u) = \int_{-\infty}^{\infty} h(u,v)dv$$

$$\int 2v\,dv$$

$$g(u) = 1;\ 0 < u < 1$$

The marginal p.d.f of v is given by:

$$g(u) = \begin{cases} 1; & 0 < u < 1 \\ 0; & \text{otherwise.} \end{cases}$$

2. The joint density function of the random variable (X, Y) is given by, f(x, y) = 8xy, 0 < x < 1, 0 < y < x = 0, elsewhere. Let us determine the marginal density of Y, conditional

density of X/Y = y and $P\left(X < \frac{1}{2}\right)$.

Solution:

Given:

$$f(x, y) = 8xy,\ 0 < x < 1,\ 0 < y < x$$

$$= 0, \text{ elsewhere}$$

Formula to be used:

$$f_{X/Y}(x/y) = \frac{f(x,y)}{f_Y(y)}$$

The joint density function of the random variable (X, Y) is:

$$f(x, y) = 8xy,\ 0 < x < 1,\ 0 < y < x$$

$$= 0, \text{ elsewhere}$$

Marginal density of Y is given by:

$$f_Y(y) = \int_y^1 8xy\,dx = 4y(1 - y^2),\ 0 < y < 1$$

Conditional density of X/Y = y

$$f_{X/Y}(x/y)=\frac{f(x,y)}{f_Y(y)}=\frac{2x}{1-y^2},0<y<x,\ 0<x<1$$

$$p(x<1/2)=\int_0^{1/2}\int_0^{x}8xy\ dy\,dx=1/16$$

3. The joint pdf of a two-dimensional RV(X, Y) is given by:

$$f(x,y)=xy^2+\frac{x^2}{8};\ 0\le x\le 2, 0\le y\le 1.$$

Let us compute:

1. $P\left(X>1/Y<\frac{1}{2}\right)$
2. $P\left(Y<\frac{1}{2}/X>1\right)$
3. $P(X<Y)$
4. $P(X+Y\le 1)$

Solution:

Given:

$$f(x,y)=xy^2+\frac{x^2}{8};\ 0\le x\le 2,\ 0\le y\le 1.$$

$$1.\ P(X>1/Y<1/2)=\frac{P\left(X>1, Y<\frac{1}{2}\right)}{P(Y<1/2)}$$

$$=\frac{\int_0^{1/2}\int_1^{2}f(x,y)dx\ dy}{\int_0^{1/2}\int_0^{2}f(x,y)dx\ dy}$$

$$=\frac{5/24}{1/4}=\frac{5}{6}$$

2. $$P\left(Y<\frac{1}{2}Px>1\right)=\frac{P\left(Y<\frac{1}{2},X>1\right)}{P(X>1)}$$

$$=\frac{5/24}{19/24}=\frac{5}{19}$$

3. $$P(X<Y)=\int_0^1\int_0^y f(x,y)dx\ dy$$

$$=\int_0^1\int_0^y\left(xy^2+\frac{x^2}{8}\right)dx\ dy$$

$$=\frac{53}{480}$$

4. $$P(X+Y\le 1)=\int_0^1\int_0^{1-y} f(x,y)dx\ dy$$

$$=\int_0^1\int_0^{1-y}\left(xy^2+\frac{x^2}{8}\right)$$

$$=\frac{13}{480}$$

1.1.3 Distribution Functions

Distribution Function of the Random Variable X or Cumulative Distribution of the Random Variable X

The distribution function of a random variable X defined in $(-\infty, \infty)$ is given by:

$$F(x)=P(X\le x)=P\{s : X(s) \le x\}$$

Note:

Let the random variable X takes values $x_1, x_2, \ldots, x_n$ with probabilities $P_1, P_2, \ldots, P_n$ and let $x_1 < x_2 < \ldots < x_n$. Then we have:

$$F(x)=P(X < x_1)=0,\ -\infty < x < x$$

$$F(x)=P(X < x_1)=0,\ P(X < x_1)+ P(X = x_1)=0 + p_1 = p_1$$

$$F(x) = P(X < x_2) = 0,\ P(X < x_1) + P(X = x_1) + P(X = x_2) = p_1 + p_2$$

$$F(x) = P(X < x_n) = P(X < x_1) + P(X = x_1) + \ldots.. + P(X = x_n)$$

$$= p_1 + p_2 + \ldots\ldots.. + p_n = 1$$

Distribution Functions

Definition

Suppose that X is a real valued random variable. The distribution function of X is the function F given by:

$$F(x) = P\ (X \le x), \quad x \in \mathbb{R}$$

The distribution function is important because it makes sense for any type of random variable, regardless of whether the distribution is discrete, continuous or even mixed and because it completely determines the distribution of X.

In the picture below, the light shading is intended to represent a continuous distribution of probability, while the darker dots represents points of positive probability F(x) is the total probability mass to the left of x.

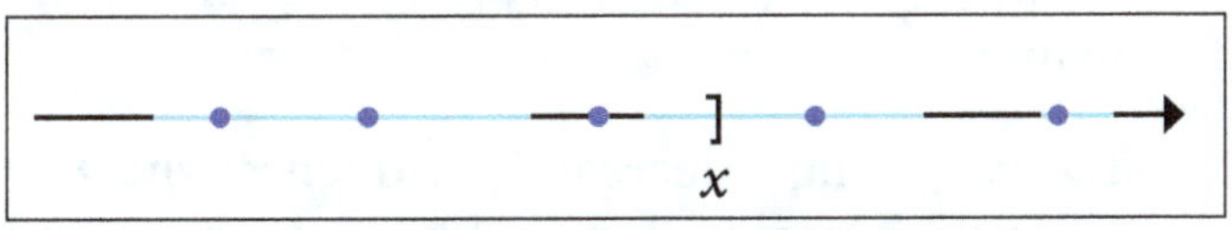

F(x) is the total probability to the left of (and including) x.

Properties of Distribution Functions

$$\text{Property}: 1\ P(a < X \le b) = F(b) - F(a),\ \text{where } F(x) = P(X \le x)$$

$$\text{Property}: 2\ P(a \le X \le b) = P(X = a) + F(b) - F(a)$$

$$\text{Property}: 3\ P(a < X < b) = P(a < X \le b) - P(X = b)$$

$$= F(b) - F(a) - P(X = b)$$

Basic Properties

A few basic properties completely characterize distribution functions. Notationally, it will be helpful to abbreviate various limits of F as follows:

$$F(x^+) = \lim_{t \downarrow x} F(t),$$

$$F(x^-) = \lim_{t \uparrow x} F(t),$$

$$F(\infty) = \lim_{t \to \infty} F(t),$$

$$F(-\infty) = \lim_{t \to -\infty} F(t)$$

1. Suppose that F is the Distribution Function of a Real Valued Random Variable X.

   F is increasing: if $x \le y$ then $F(x) \le F(y)$.

   $F(x^+) = F(x)$ for $x \in \mathbb{R}$. Thus, F is continuous from the right.

   $F(x^-) = P(X < x)$ for $x \in \mathbb{R}$. Thus, F has limits from the left.

   $F(-\infty) = 0$.

   $F(\infty) = 1$.

Proof:

- If x≤y, then $\{X \le x\} \subseteq \{X \le y\}$. Hence, the result follows from the increasing property of probability.
- Fix x∈R. Let x1 > x2 > ... be a decreasing sequence with $x_n \to x$ as $n \to \infty$. The intervals $(-\infty, x_n)$ are decreasing in n and have intersection $(-\infty, x)$. The result now follows from the continuity theorem for decreasing events.
- Fix x∈R. Let $x_1 < x_2 <$ ... be an increasing sequence with $x_n \to x$ as $n \to \infty$. The intervals $(-\infty, x_n)$ are increasing in 'n' and have union $(-\infty, x)$. The result now follows from the continuity theorem for increasing events.
- Let $x_1 > x_2 >$ ... be a decreasing sequence with $x_n \to -\infty$ as $n \to \infty$. the intervals $(-\infty, x_n)$ are decreasing in 'n' and have intersection R. The result now follows from the continuity theorem for decreasing events.
- Let $x_1 < x_2 <$ ... be an increasing sequence with $x_n \to \infty$ as $n \to \infty$. The intervals $(-\infty, x_n)$ are increasing in n and have union R. The result now follows from the continuity theorem for increasing events.

The following result shows how the distribution function can be used to compute the probability that X is in an interval. That a probability distribution on R is completely determined by the probabilities of intervals thus, the distribution function determines the distribution of X.

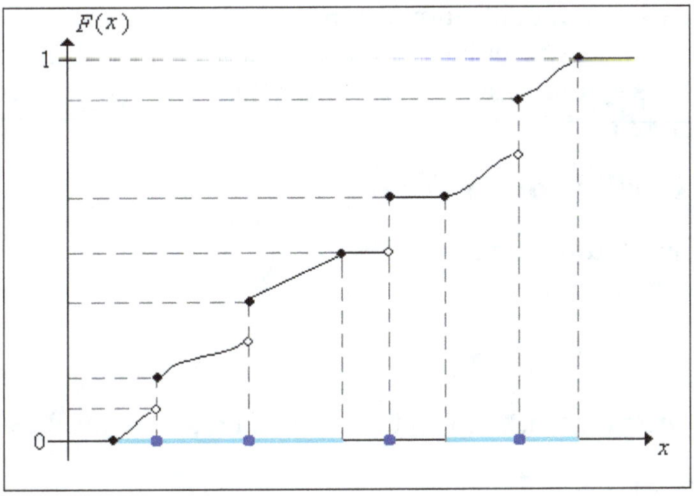

The graph of a distribution function.

2. Suppose that F is the Distribution Function of a Real Valued Random Variable X. If a, b∈R and a<b, then:

$$P(X=a)=F(a)-F(a^-)$$

$$P(a<X\le b)=F(a)-F(a)$$

$$P(a<X<b)=F(b^-)-F(a)$$

$$P(a\le X\le b)=F(b)-F(a^-)$$

$$P(a\le X<b)=F(b^-)-F(a^-)$$

Proof:

These results follow from the definition, the basic properties and the difference rule:

$P(B|A)=P(B)-P(A)$ if A,B are events and A⊆B.

Conversely, if a Function F on R satisfies the basic properties, then the probability distribution on R with F as the distribution function.

If X has a continuous distribution, then by definition:

$P(X=x)=0$ so $P(X<x)=P(X\le x)$ for $x\in R$

Hence from the above theorems:

$$F(x^-)=F(x^+)=F(x)$$

If X has a continuous distribution, then the distribution function F is continuous.

Thus, the two meanings of continuous come together, continuous distribution and continuous function in the calculus sense. Next that the distribution of a real valued random variable X is symmetric about a point a⊆R if the distribution of X−a is same as the distribution of a−X.

Cumulative Distribution Function

The cumulative distribution function (cdf) is the probability that the variable takes a value less than or equal to x. That is:

$$F(x) = Pr[X \leq x] = \alpha$$

For a continuous distribution, this can be expressed mathematically as:

$$F(x) = \int x - \infty f(\mu) d\mu$$

For a discrete distribution the cdf can be expressed as:

$$F(x) = \sum xi = of(i)$$

The following is the plot of the normal cumulative distribution function.

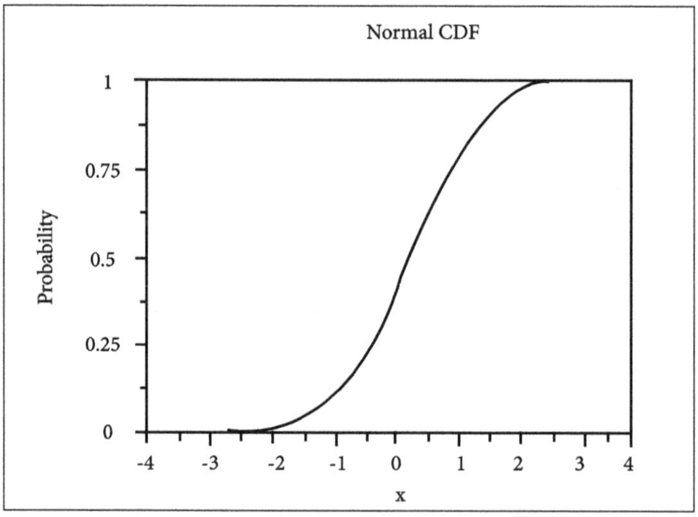

The horizontal axis is the allowable domain for the given probability function. Since the vertical axis is a probability, it must fall between zero and one. It increases from zero to one as we go from left to right on the horizontal axis.

Properties of CDF of A R.V. 'X':

(i) $0 \leq F(x) \leq 1, -\infty < x < \infty$

(ii) $\underset{x \to \infty}{Lt} F(x) = 0, \quad \underset{x \to -\infty}{Lt} F(x) = 1$

(iii) $P(a \le X \le b) = \int_a^b f(x)dx = F(b) - F(a)$

(iv) $F'(x) = \frac{dF(x)}{dx} = f(x) \ge 0$

(v) $P(X = x_i) = F(x_i) - F(x_i - 1)$

Example:

In the experiment of throwing a coin twice, the sample space S is $S = \{HH, HT, TH, TT\}$. Let X be a random variable chosen such that X(S) = x (the number of heads).

Note:

Any random variable whose only possible values are 0 and 1 is called as Bernoulli random variable.

1.2 Binomial and Poison Distributions

Binomial

A random variable X is said to follow binomial distribution if its probability law is given by,

$$P(x) = p(X = x \text{ successes}) = nC_x p_x q^{n-x}$$

Where $x = 0, 1, 2, \ldots\ldots, n, \ p + q = 1$

Note

Assumptions in Binomial Distribution

- There are only two possible outcomes for each trail (success or failure).
- The probability of a success is the same for each trail.
- There are 'n' trails, where 'n' is a constant.
- The 'n' trails are independent.

Moment Generating Function (MGF) of a Binomial Distribution about Origin

$$M_X(t) = \sum_{x=0}^{n} e^{tx} p(x)$$

Let 'X' be a random variable which follows binomial distribution then MGF about origin is given by:

$$E\left[e^{tX}\right] = M_x(t) = \sum_{x=0}^{n} e^{tx} p(x)$$

$$= \sum_{x=0}^{n} e^{tx} nC_x p^x q^{n-x} \qquad \left[\because p(x) = nC_x p^x q^{n-x}\right]$$

$$= \sum_{x=0}^{n} \left(e^{tx}\right) p^x nC_x q^{n-x}$$

$$= \sum_{x=0}^{n} \left(pe^t\right)^x nC_x q^{n-x}$$

$$\therefore M_x(t) = \left(q + pe^t\right)^n$$

Mean and Variance of Binomial Distribution

$$M_x(t) = \left(q + pe^t\right)^n$$

$$\therefore M_x' = n\left(q + pe^t\right)^{n-1} \cdot pe^t$$

Put $t = 0$, we get,

$$M'_x(0) = n(q+p)^{n-1} \cdot p$$

$$\text{Mean} = E(X) = np \qquad \left[\because (q+p) = 1\right] \left[\text{Mean } M_X'(0)\right]$$

$$M_x''(t) = np\left[\left(q + pe^t\right)^{n-1} \cdot e^t + e^t(n-1)\left(q + pe^t\right)^{n-2} \cdot pe^t\right]$$

Put $t = 0$, we get,

$$M_x''(t) = np\left[(q+p)^{n-1} + (n-1)(q+p)^{n-2} \cdot p\right]$$

$$= np\left[1 + (n - 1p)\right]$$

$$= np + n^2p^2 - np^2$$

$$= n^2p^2 + np(1-p)$$

$$M_x''(o)=n^2p^2+npq \qquad [\because 1-p=q]$$

$$M_x''(o)=E(X^2)=n^2p^2+npq$$

$$\text{Var}(X)==E(X^2)-[E(X)]^2=n^2/p^2+npq-n^2/p^2=npq$$

$$\text{Var}(X)=npq$$

$$S.D=\sqrt{npq}$$

Additive Property of Binomial Distribution

Theorem

The sum of two binomial variants is not a binomial variate.

Proof:

Let X and Y be two independent binomial variates with parameter (n1, p1) and (n2, p2) respectively.

Then:

$$M_x(t)=(q_1+p_1e^t)^{n_1}, \; M_Y(t)=(q_2+p_2e^t)^{n_2}$$

$$\therefore M_{X+Y}(t)=M_X(t)\cdot M_Y(t) \qquad [\because X\&Y \text{ are independent R.V.'s}]$$

$$=(q_1+p_1e^t)^{n_1}\cdot(q_2+p_2e^t)^{n_2}$$

RHS cannot be expressed in the form $(q+pe^t)^n$. Hence by uniqueness theorem of MGF, X+Y is not a binomial variate. Hence in general, the sum of two binomial variates is not a binomial variate.

Poisson

A random variable X is said to follow Poisson distribution, if its probability law is given by:

$$P(X=x)=p(x)=\frac{e^{-\lambda}\lambda^x}{X!}, \quad x=0,1,2,\ldots\ldots\infty$$

Poisson distribution is a limiting case of binomial distribution under the following conditions or assumptions:

- The number of trails 'n' should be infinitely large i.e. $n\to\infty$.

- The probability of successes 'p' for each trail is infinitely small.
- np = λ should be finite, where λ is a constant.

To find MGF:

$$MX(t)=E\left(e^{tx}\right)$$

$$=\sum_{x=0}^{\infty} e^{tx}\, p(x)$$

$$=\sum_{x=0}^{\infty} e^{tx}\left(\frac{\lambda^{x} e^{\lambda}}{x!}\right)$$

$$=\sum_{x=0}^{\infty} \frac{e^{-\lambda}\left(\lambda e^{t}\right)^{x}}{x!}$$

$$=e^{-\lambda} \sum_{x=0}^{\infty} \frac{\left(\lambda e^{t}\right)^{x}}{x!}$$

$$=e^{-\lambda}\left[1+\lambda e^{t}+\frac{\left(\lambda e^{t}\right)^{2}}{2!}+\ldots..\right]$$

$$=e^{-\lambda}\, e^{\lambda e^{t}}=e^{\lambda\left(e^{t}-1\right)}$$

Hence, $M_X(t)=e^{\lambda\left(e^{t}-1\right)}$

To find the mean and variance:

$$M_X(t)=e^{\lambda\left(e^{t}-1\right)}$$

$$\therefore\ M_X'(t)=e^{\lambda\left(e^{t}-1\right)}\cdot e^{t}$$

$$M_X'(0)=e^{-\lambda}\cdot\lambda$$

$$\mu_1=E(X)=\sum_{x-0}^{\infty} x\cdot p(x)$$

$$=\sum_{x-0}^{\infty} x\cdot\frac{e^{-\lambda}\lambda^{x}}{x!}=\sum_{x-0}^{\infty} \frac{x\cdot e^{-\lambda}\,\lambda\,\lambda^{x-1}}{x!}$$

$$=0+e^{-\lambda}.\lambda\sum_{x-1}^{\infty}\frac{x.\lambda^{x-1}}{x!}$$

$$=\lambda e^{-\lambda}\cdot\sum_{x-1}^{\infty}\frac{\lambda^{x-1}}{(x-1)!}$$

$$=\lambda e^{-\lambda}\left[1+\lambda+\frac{\lambda^2}{2!}+.........\right]$$

$$=\lambda e^{-\lambda}\cdot e^{\lambda}$$

Mean $=\lambda$

$$\mu_2=E\left[X^2\right]=\sum_{x-0}^{\infty}x^2\cdot p(x)=\sum_{x-0}^{\infty}x^2\cdot\frac{e^{-\lambda}\lambda^x}{x!}$$

$$=\sum_{x-0}^{\infty}\{x(x-1)+x\}\cdot\frac{e^{-\lambda}\lambda^x}{x!}$$

$$=\sum_{x-0}^{\infty}\frac{x\,(x-1)e^{-\lambda}\lambda^x}{x!}+\sum_{x-0}^{\infty}\frac{x.e^{-\lambda}\lambda^x}{x!}$$

$$=e^{-\lambda}\,\lambda^2\sum_{x-0}^{\infty}\frac{\lambda^{x-2}}{(x-2)(x-3).......1}+\lambda$$

$$=e^{-\lambda}\,\lambda^2\sum_{x-0}^{\infty}\frac{\lambda^{x-2}}{(x-2)!}+\lambda$$

$$=e^{-\lambda}\lambda^2\left[1+\frac{\lambda}{1!}+\frac{\lambda^2}{2!}+........\right]+\lambda$$

Variance, $\mu_2=E\left(X^2\right)-\left[E(X)\right]^2=\lambda^2+\lambda-\lambda^2=\lambda$

Variance $=\lambda$

Hence, Mean = Variance $=\lambda$

Problems

1. In a binomial distribution, the mean is 4 and variance is 3, let us determine P(X=0).

Solution:

Given:

$$np = 4$$

$$npq = 3$$

Formula to be used:

$$p = 1 - q$$

$$P(X = x) = nc_x p^x q^{n-x}$$

Hence,

$$q = 3/4$$

$$p = 1 - q = 1/4$$

Since, $np = 4$, $n = 16$.

$$P(X = x) = nc_x p^x q^{n-x}$$

$$= 16c_x \left(\frac{1}{4}\right)^x \left(\frac{3}{4}\right)^{16-x} \quad x = 0,1,2....$$

$$P(X = 0) = \left(\frac{3}{4}\right)^{16} = 0.01002$$

2. If X is binomially distributed with n = 6 such that P(X = 2) = 9P(X = 4), let us determine E(x) and Var (x).

Solution:

Given:

$$n = 6$$

$$P(X = 2) = 9P(X = 4)$$

$$6C_2\ p^2q^4 = 9\ (6C_4p^4q^2)$$

$$q = 3p$$

$$p = 1/4$$

$$E(X) = 1.5$$

$$Var(X) = 9/8$$

3. Suppose that in a class 5% girls are shorter than 5ft. Let us know how many girls should be present in each class in order that the probability of atleast one girl shorter than 5ft is 50% or more.

Solution:

X = No. of girls who are shorter than 5ft.

$$X \sim Bin(n, 0.5)$$

$$P(X \geq 1) = 1 - P(X = 0) = 1 - (0.95)^n$$

$$\therefore 1 - (0.95)^n \geq \frac{1}{2}$$

$$\therefore n \geq 14$$

4. The tissues produced by a company are defective with probability 0.01 and are independent of each other. The company sells the tissues papers in packs of size 10 and offers a money-back guarantee if more than one of the 10 tissue paper packs in the pack is found to be defective. If we buy 3 packs, let us know what is the probability that at most one pack will be returned.

Solution:

P (returning a packets) = $P(X > 1)$

$$P(X > 1) = 1 - P(X = 0) - P(X = 1)$$

$$\therefore P(X > 1) = 1 - (0.99)^{10} - 10(0.99)^9(0.01)$$

$$\therefore P(X > 1) = 0.04$$

Now let Y denote the number of packs returned.

Then $Y \sim Bin\ (3, 0.04)$

$$P(Y = 0) + P(Y = 1) = {}^3C_0(0.04)^0(0.96)^3 + {}^3C_1(0.04)(0.96)^2$$

Hence, $P = 0.995328$

5. A binomial variable on 80 trials has 6 as its std deviation. Let us know whether this statement is valid or not.

Solution:

$$Var\ (X) = np\ (1 - p) = n(p - p2)$$

$n(p-p^2)=\phi(p)$ say

maximum value of $\phi(p) \rightarrow \phi'(p)=0$

$1-2p=0$

$\therefore p=\frac{1}{2}$

$\phi''(p)=-2n<0$

Hence, p= ½ is the point of maximum of Variance.

$$\text{Variance}_{max}=n\left(\frac{1}{2}\right)\left(\frac{1}{2}\right)=\frac{n}{2}$$

Here variance= (6)2=36 which is greater than (80/4)

Hence, the statement is valid.

6. A machine produces 5% defective items. 12 items are inspected every hour. The machine has to be shut down if more than 2 defectives are found. Let us know what is the probability that the machine has to be shut for the first time at the time of the 3rd inspection.

Solution:

Let X denote the number of defective items.

$$X \sim \text{Bin}(12, 0.05)$$

$$P(\text{shutting down}) = P(X \geq 2)$$

$$=1-P(X<2)=1-[P(X=0)+P(X=1)]$$

$$P(\text{shutting down})=1-\left[(0.95)^{12}+{}^{12}C_1(0.05)(0.95)^{11}\right]$$

$$\therefore P(\text{shutting down})=0.1183$$

If Y= Number of failures,

Then if machine has to be shut down for the first time at the 4th inspection. We need to find P(Y=2) and Y follows geometric distribution.

$$P(Y=)=p(1-p)^2=(0.1183)(0.8816)^2$$

$$P(Y=2)=0.0919$$

7. Ayush is given questions in a Mathematics exam one by one until he fails to answer correctly. The probability that Ayush answers a question correctly is m. The probability that he will quit after answering an odd number of questions is 0.9.Let us find the value of m.

Solution:

Let X be the number of questions asked to Ayush. Then X follows a geometric distribution.

He answers the last question wrongly. Hence,

$$P(X=k)=m^{k-1}(1-m)$$

Given that the probability of Ayush answering an odd number of questions= 0.9

$$\therefore \sum_{k=0}^{\infty} P(X=2k+1)=0.9$$

Let $(1-m)=n$

$$\therefore \sum_{k=0}^{\infty} n m^{2k}=0.9$$

$$\therefore \frac{n}{1-m^2}=0.9$$

Hence,

$$m=\frac{1}{9}$$

8. If the probability of a defective bulb is 0.1; then let us find the moment of skewness and kurtosis for 400 bulbs following Binomial distribution.

Solution:

$$\mu_r' = \frac{d^t M_x(t)}{dt^r}, \text{at } t=0$$

$$\therefore \mu_3 = \frac{d^3 M_x(t)}{dt^3}, \text{at } t=0$$

$$\therefore \mu_3 = np(1-p)(1-2p)$$

$$\beta = \frac{\mu_3}{\sigma^3} = \frac{np(1-p)(1-2p)}{(npq)^{3/2}} = \frac{1-2p}{\sqrt{npq}}$$

$$\beta = \frac{1-2p}{\sqrt{npq}} = \frac{0.8}{\sqrt{(400)(0.1)(0.9)}}$$

$$\therefore \beta = 0.134$$

$$\mu_r' = \frac{d'M_x(t)}{dt^t}, \text{at } t = 0$$

$$\therefore \mu = \frac{d\ M\ (t)}{dt}, \text{at } t = 0$$

$$\gamma = \frac{\mu_4}{\sigma^4} - 3 = \frac{1-6pq}{npq}$$

$$\gamma = \frac{1-6pq}{npq} = \frac{1-6(0.1)(0.9)}{400(0.1)(0.9)}$$

$$\therefore \gamma = 0.013$$

Since moment of skewness =0.134 distribution is positively skewed. And since moment of kurtosis > 0, distribution is platykurtic.

9. X (t) be a Poisson process with arrival rate λ. Let us find E{(X(t) - X(s)2} for t > s.

Solution:

Given:

X (t) is a Poisson process with rate λ.

Formula to be used:

$$EX(t) = \lambda t$$

$$EX^2(t) = (\lambda t)^2 + \lambda t$$

$$\text{and, Var } X(t) = \lambda t.$$

Now

$$E(X(t) - X(s))^2 = E\{X^2(t) + X^2(s) - 2X(t)X(s)\}$$

$$= EX^2(t) + EX^2(s) - 2\,EX(t)X(s)$$

$$= (\lambda t)^2 + \lambda t + (\lambda S)^2 + \lambda S - 2\,E(X(t) - X(s))X(s) + X^2(s)$$

$$= (\lambda t)^2 + \lambda t + (\lambda S)^2 + \lambda S - 2\,E\{X(s)\}\ E\{(X(t) - X(s))\} - 2\,E\{X^2(s)\}$$

$(\therefore\ X(t)$ is a process of independent increments $t > s)$

$$= (\lambda t)^2 + \lambda t + (\lambda S)^2 + \lambda S - 2(\lambda S)\ (\lambda t - \lambda S) - 2\left[(\lambda S)^2 \_ \lambda S\right]$$

$$= (\lambda t)^2 + \lambda t + (\lambda S)^2 + \lambda S - 2(\lambda S)(\lambda t) + 2(\lambda S)^2 - 2(\lambda S)^2 - 2(\lambda S)$$

$$= (\lambda t)^2 + (\lambda s)^2 + \lambda t - \lambda s - 2(\lambda s)(\lambda t)$$

$$= \lambda^2\ (t^2 + s^2 - 2st) + \lambda(t - s)$$

$$= \lambda^2 (t - s)^2 + \lambda(t - s)$$

$E\left\{(X(t) - X(s)^2\right\} = (t - s)^2 \lambda^2 + (t - s)\lambda$ for $t > s$

10. If X and Y are independent Poisson variates, let us determine the conditional distribution of X given X + Y = n.

Solution:

Given:

X and Y are independent Poisson variates with parameters λ and μ respectively. Then, X + Y is also a Poisson variate with parameter λ + μ.

Formula to be used:

$$P[X = r / X + y = n] = \frac{P[X = r / X + y = n]}{P(X + y = n)}$$

$$= \frac{P[X = r]P[y = n - r]}{P[X + y = n]} = \frac{\dfrac{e^{-\lambda}.\lambda^r e^{-\mu}\ \mu^{n-r}}{r!(n-r)!}}{\dfrac{e^{-\lambda+\mu}.(\lambda+\mu)^n}{n!}}$$

$$=\frac{n!}{r!(n-r)!}\left(\frac{\lambda}{\lambda+\mu}\right)^{r}\left(\frac{\mu}{\lambda+\mu}\right)^{n-r}$$

$$=n_{cr}\,p^{r}q^{n-r} \text{ where } P=\frac{\lambda}{\lambda+\mu},\ q=\frac{\mu}{\lambda+\mu}$$

Hence, the conditional distribution of X given X+Y = n is a binomial distribution with parameters n and,

$$P=\frac{\lambda}{\lambda+\mu};q=1-p$$

11. Let us derive the probability mass function of Poisson distribution as a limiting case of Binomial distribution.

Solution:

We know that the Binomial distribution is,

$$P(X=x)\ =nC_{x}p^{x}q^{n-x}$$

$$=\frac{n!}{(n-x)!\,x!}p^{x}(1-p)^{n-x}$$

$$=\frac{1.2.3.............(n-x)(n-x+1)....np^{n}}{1.2.3....(n-x)x!}\frac{(1-p)^{n}}{(1-p)^{x}}$$

$$=\frac{1.2.3....(n-x)(n-x+1)....n}{1.2.3....(n-x)x!}\left(\frac{p}{1-p}\right)^{x}(1-p)^{n}$$

$$=\frac{n(n-1)(n-2)....(n-x+1)}{x!}\frac{\lambda^{x}}{n^{x}}\frac{1}{\left(1-\frac{\lambda}{n}\right)^{x}}\left(1-\frac{\lambda}{n}\right)^{n}$$

$$=\frac{n(n-1)(n-2)......n-x+1}{x!}x\left(1-\frac{\lambda}{n}\right)^{n}\left(1-\frac{\lambda}{n}\right)^{-x}$$

$$P(X=x)=\frac{1\left(1-\frac{1}{n}\right)\left(1-\frac{2}{n}\right).........\left\{1-\left(\frac{x-1}{n}\right)\right\}}{x!}\lambda^{x}\left(1-\frac{\lambda}{n}\right)^{n-x}$$

$$=\frac{\lambda^{x}}{x!}1\left(1-\frac{1}{n}\right)\left(1-\frac{2}{n}\right).........\left\{1-\left(\frac{x-1}{n}\right)\right\}\left(1-\frac{\lambda}{n}\right)^{n-x}$$

When $n \to \infty$,

$$P(X=x)=\frac{\lambda^x}{x!}\underset{n\to\infty}{\text{lt}}\left[1\left(1-\frac{1}{n}\right)\left(1-\frac{2}{n}\right)\cdots\cdots\left\{1-\left(\frac{x-1}{n}\right)\right\}\left(1-\frac{\lambda}{n}\right)^{n-x}\right]$$

$$=\frac{\lambda^x}{x!}\underset{n\to\infty}{\text{lt}}\left(1-\frac{1}{n}\right)\underset{n\to\infty}{\text{lt}}\left(1-\frac{2}{n}\right)\cdots\cdots\underset{n\to\infty}{\text{lt}}\ 1-\left(\frac{x-1}{n}\right)$$

We know that,

$$\underset{n\to\infty}{\text{lt}}\left(1-\frac{\lambda}{n}\right)^{n-x}=e^{-\lambda}$$

and, $\underset{n\to\infty}{\text{lt}}\left(1-\frac{1}{n}\right)=\underset{n\to\infty}{\text{lt}}\left(1-\frac{2}{n}\right)\cdots=\underset{n\to\infty}{\text{lt}}\left(-\left(\frac{x-1}{n}\right)\right)=1$

$$\therefore\ P(X=x)=\frac{\lambda^x}{x!}e^{-\lambda}, x=0,1,2,\ldots\ldots\infty$$

1.2.1 Normal Distribution

Moment Generating Functions of a Normal Distribution

MGF is given by:

$$m_x(t)=\int_{-\infty}^{+\infty}e^{tx}f(x)\,dx=\frac{1}{\sigma\sqrt{2\pi}}\int_{-\infty}^{+\infty}e^{tx}\,e^{\frac{-(x-\mu)}{2\sigma^2}}\,dx$$

$$=\frac{1}{\sqrt{2\pi}}\int_{-\infty}^{+\infty}e^{(tx(\mu+\sigma z))}\,e^{-z2/2}dz$$

where $z=\frac{x-\mu}{\sigma}$,

$$=e^{\mu t}\frac{1}{\sqrt{2\pi}}\int_{-\infty}^{+\infty}\left(e^{-\frac{1}{2}\left(z^2-2+\sigma z\right)}\right)dz$$

$$=e^{\mu t}\,e^{t^2\sigma^2/2}$$

$$=e^{\mu t}+\frac{t^2\sigma^2}{2}\cdot\frac{1}{\sqrt{2\pi}}\int_{-\infty}^{+\infty}e^{-\frac{1}{2}\left(z^2-\sigma z\right)^2}dz$$

$$=e^{\mu t}+\frac{t^2\sigma^2}{2}\cdot\frac{1}{\sqrt{2\pi}}\cdot\sqrt{2\pi}$$

$$=e^{\mu t+\frac{t^2\sigma^2}{2}}$$

Problems

1. X is normally distributed and the mean of X is 12 and S.D. is 4. Let us also see the probability of the following:

- $X \geq 20$
- $0 \leq X \leq 12$
- And x′, when P(X>x’) = 0.24
- Get value also for x_0^1 and x , when $P\left(x_0^1<X<x_1^1\right)=0.50$ and $P\left(X>x_1^1=0.25\right)$.

Solution:

Given:

Mean of X = 12

S.D = 4

Formula to be used:

$$Z=\frac{X-\mu}{\sigma}$$

1. $P(x\geq 20)=P(\geq \geq 2),\ Z=\frac{X-\mu}{\sigma}$

$$=0.5-P(0\leq Z\leq 2)$$

$$=0.5-0.4772=0.0228$$

2. $P(0\leq x\leq 12)=P(-3\leq Z\leq 0)$

$$=P(0\leq Z\leq 3)$$

$$=0.49865$$

3. $P\left(x>x^1\right)=P\left(Z>Z_1\right)=0.24$

$$Z = \frac{x^1 - 12}{4} = Z_1 \text{ (say)}$$

$$\Rightarrow P(0 < Z < Z_1) = 0.26$$

$$\Rightarrow Z_2 = 0.71$$

$$\text{Hence } \frac{x^1 - 12}{4} = 0.71$$

$$\Rightarrow x^1 = 14.84$$

4. $P(x_0^1 < x < x_1^1) = 0.50$ and $P(x > x_1^1) = 0.25$

When $x = x_1^2$, $Z = x_1^1 - 12 = \overline{4} = Z_1$

$$\Rightarrow P(Z > Z_1) = 0.25$$

When $X = x_0^1$, $Z - \frac{x_0^1 - 12}{4} = -Z_1$

$$\Rightarrow P(0 < Z < Z_1) = 0.25$$

$$\Rightarrow Z_1 = 0.67$$

$$\therefore x_1^1 = 14.68$$

$$x_0^1 = 9.32$$

1.2.2 Related Properties

Properties of Gaussian Random Variables:

- The Gaussian random variables are completely defined by their means, variances and covariances.
- If the Gaussian random variables are uncorrelated, then they are statistically independent.
- All marginal density functions derived from N-variate Gaussian density functions are Gaussian.
- All conditional density functions are also Gaussian.
- All linear transformations of Gaussian random variables are also Gaussian.

$$\rho_{XY}(t_1,t_2)=\frac{c_{XY}(t_1,t_2)}{\sqrt{Var(X(t_1))\ Var(X(t_2))}}$$

Let $\{X(t)\}$ and $\{Y(t)\}$ be two random. Then the cross correlation between them is also defined as:

- If the processes $\{X(t)\}$and $\{Y(t)\}$ are jointly wide-sense stationary, then
  - $E[X(t) \bullet Y(t-\tau)]$ is a function of τ, denoted by $Rxy(\tau)$. This function $Rxy(\tau)$ is called as cross correlation function of the processes X(t) and Y(t).
- If $R_x y(\tau) = \mu_x \times \mu_y$, then the processes X (t) and Y (t) are independent.

2

Test of Hypothesis

2.1 Population and Sample

Population

It is the collection of objects. A population may be finite or infinite according to the number of objects in the population.

For example: the number of students in a college is a finite population.

Samples

A population commonly contains several individuals to study conveniently, so an investigation is often restricted to one or more samples drawn from it. A well-chosen sample will have most of the information about a particular population parameter but the relation between the sample and the population must be such as to permit the true inferences to be made about a population from that sample.

The important difference between a population and sample has to do with how observations are assigned to the data set.

A population has all of the elements from a set of data.

A sample contains one or more observations from the population.

Depending on the sampling method, a sample can have fewer observations than the population, the same number of observations or more observations. More than one sample can be derived from the same population.

Other differences have to do with notation, nomenclature and computations. For example,

A a measurable characteristic of a population, such as a mean or standard deviation, is called a parameter but a measurable characteristic of a sample is known as a statistic.

The mean of a population is denoted by the symbol μ but the mean of a sample is referred by the symbol x.

The formula for the standard deviation of a population is varied from the formula for the standard deviation of a sample.

A sampling method is a procedure for choosing the sample elements from a population. Simple random sampling defines to a sampling method that has the following properties.

- The sample consists of n objects.
- All possible samples of n objects are equally likely to occur.
- The population has N objects.

An important advantage of simple random sampling is that it allows researchers to use statistical methods to analyze sample results. For example, provided a simple random sample, researchers can use statistical methods to define a confidence interval around a sample mean. Statistical analysis is not appropriate when the non-random sampling methods are used.

There are several methods to get a simple random sample. One way would be the lottery method. Each of the N population members is assigned a unique number. The numbers are placed in a bowl and thoroughly mixed. Then, a blind-folded researcher selects n numbers. Population members having the selected numbers are included in the sample.

Sampling with Replacement and Without Replacement

Suppose we use the lottery method described above to select a simple random sample. After we pick a number from the bowl, we can put the number aside or we can put it back into the bowl. If we put the number back in the bowl, it might be selected more than once; if we put it aside, it can selected only one time.

When a population element can be selected more than one time, we are sampling with replacement. When a population element can be chose only one time, we are sampling without replacement.

Unbiasedness and Precision

The terms unbiased and the precision have acquired special meanings in statistics. When we say that a measurement is unbiased we mean that the average of a large set of unbiased measurements will be close to the true value. When we say it is precise we mean that it is repeatable. Repeated measurements will be close to one another, but not importantly close to the true value.

Variation Between Samples

Even if we check that every member of a population has a known and usually an equal, chance of being included in a sample, it does not follow that a series of samples drawn from one population and fulfilling this criterion will be identical. They will represent the chance variations from one to another and the variation might be slight or considerable. For example, a series of samples of the body temperature of healthy people would represent very little variation from one to another, but the variation between samples of the systolic blood pressure would be considerable. Thus the variation between samples depends partly on the amount of variation in the population from which they are drawn.

Example of a Population

Let us consider five project engineers in department:

Total experience of (X) 2, 3, 6, 8, 11 years

The company performing statistical report.

Employees expertise based on experience survey should include:

- Average experience.
- Variance.
- Standard deviation.

Mean of Populations

Average experience mean:

$$\mu = \frac{2+3+6+8+11}{5} \frac{30}{5} = 6 \text{ years}$$

Variance of Population:

$$\sigma^2 = \sum \frac{(x_i - \mu)^2}{n}$$

$$\sigma^2 = \frac{(2-6)^2 + (3-6)^2 + (6-6)^2 + (8-6)^2 + (11-6)^2}{5}$$

$$\sigma^2 = \frac{16+9+0+4+25}{5} = 10.8$$

Standard Deviation of Population:

$$\text{s.d.} = \sqrt{\sigma^2} = \sigma$$

$$\text{s.d.} = \sqrt{10.8} = \sigma$$

$$\sigma = \sqrt{10.8} = 3.29$$

Sampling Example

Manager assigns engineers at random,

Every time chooses the first engineer she sees,

Same engineer could do both,

Let's say she picks (2, 2)

Mean of sample $\bar{X} = (2+2)/2 = 2$

We want to make inferences about true μ.

Samples of 2

Replacement she will go to project department twice,

Pick engineer randomly,

Potentially 25 possible teams,

25 samples of size two,

$$5 * 5 = 25$$

Order matters (6, 11) are different from (11, 6).

Population of Samples

All possible combinations are:

$$(2, 2)(2, 3)(2, 6)(2, 8)(2, 11)$$

$$(3, 2)(3, 3)(3, 6)(3, 8)(3, 11)$$

$$(6, 2)(6, 3)(6, 6)(6, 8)(6, 11)$$

$$(8, 2)(8, 3)(8, 6)(8, 8)(8, 11)$$

$$(11, 2)(11, 3)(11, 6)(11, 8)(11, 11)$$

Population of Averages

Average experience or sample means are: $\bar{X}_1$

$$(2)(2.5)(3)(5)(6.5)$$

$$(2.5)(3)(4.5)(5.5)(7)$$

$(3)(4.5)(6)(7)(8.5)$

$(5)(5.5)(7)(8)(9.5)$

$(6.5)(7)(8.5)(9.5)(11)$

Mean of Population Means

Mean of sampling distribution of means is:

$$\mu_x = \frac{(2)+(2.5)+(3)+(5)+...+(11)}{25} = \frac{150}{25} = 6$$

This confirms theorem that states:

$$E(\bar{X}) = \mu_{\bar{x}} = \mu = 6$$

Variance of Sample Means

Variance of sampling distribution of means $(\bar{X}_i - \bar{X})^2$:

$(2-6)^2 (2.5-6)^2 (3-6)^2 (5-6)^2 (6.5-6)^2$

$(2.5-6)^2 (3-6)^2 (4.5-6)^2 (5.5-6)^2 (7-6)^2$

$(3-6)^2 (4.5-6)^2 (6-6)^2 (7-6)^2 (8.5-6)^2$

$(5-6)^2 (5.5-6)^2 (7-6)^2 (8-6)^2 (9.5-6)^2$

$(6.5-6)^2 (7-6)^2 (8.5-6)^2 (9.5-6)^2 (11-6)^2$

Calculating Values:

(16)(12.25)(9)(1) (0.25)

(12.25) (9)(2.25)(0.25)(1)

(9) (2.25) (0) (1) (6.25)

(1) (0.25) (1) (4)(12.25)

(0.25) (1) (6.25)(12.25) (25)

Variance is:

$$\sigma_{\bar{x}}^2 = \frac{\sum(X_i - X)^2}{n} = \frac{135}{25} = 5.4$$

Therefore standard deviation is:

$$\sigma_{x} = \sqrt{5.4} = 2.32$$

These results hold for theorem:

$$\sigma_{\bar{x}}^2 = \frac{\sigma^2}{n}$$

Where n is size of samples. Then we see that:

$$\sigma_{\bar{x}}^2 = \frac{\sigma^2}{n} = \frac{10.8}{2} = 5.40$$

2.1.1 Confidence Interval of Mean from Normal Distribution

A confidence interval provides an estimated range of values which is similar to include an unknown population parameter, the estimated range being determined from a given set of sample data.

The common notation for the parameter in question is θ. Often, this parameter is the population mean μ, which is calculated through the sample mean.

The level C of a confidence interval provides the probability that the interval produced by the method employed includes the true value of the parameter $\theta\,\bar{x}$.

When we calculate a confidence interval on the mean, we compute the mean of a sample in order to estimate the mean of the population. Clearly, if we already knew the population mean, there would be no necessary for a confidence interval. However, to explain how confidence intervals are constructed, we are going to work backwards and begin by assuming characteristics of the population. Then we will show how sample data can be used to construct a confidence interval.

Let us assume that the weights of 10-year-old children are generally distributed with a mean of 90 and a standard deviation of 36. What is the sampling distribution of the mean for a sample size of 9? The sampling distribution of the mean that the mean of the sampling distribution is μ and the standard error of the mean is:

$$\sigma_M = \frac{\sigma}{\sqrt{N}}$$

For the present example, the sampling distribution of the mean has a mean of 90 and a standard deviation of 36/3 = 12. Note that the standard deviation of a sampling distribution is its standard error. The below figure represents this distribution. The shaded area shows the middle 95% of the distribution and stretches from 66.48 to 113.52. These limits were computed by adding and subtracting 1.96 standard deviations to/from the mean of 90 as follows:

$$90-(1.96)(12)=66.48$$

$$90+(1.96)(12)=113.52$$

The value of 1.96 is based on the fact that 95% of the area of a normal distribution is within 1.96 standard deviations of the mean, 12 is the standard error of the mean.

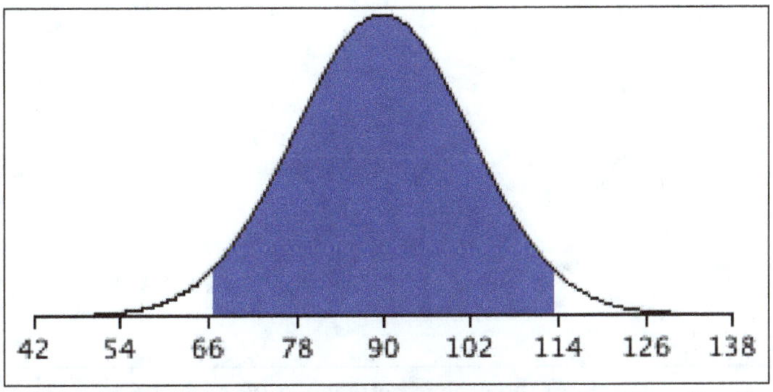

The sampling distribution of the mean for N=9. The middle 95% of the distribution is shaded.

Figure above represents that 95% of the means are no more than 23.52 units (1.96 standard deviations) from the mean of 90. Now taken the probability that a sample mean computed in a random sample is within 23.52 units of the population mean of 90. Since 95% of the distribution is within 23.52 of 90, the probability that the mean from any provided sample will be within 23.52 of 90 is 0.95. This means that if we repeatedly calculate the mean (M) from a sample and create an interval ranging from M - 23.52 to M + 23.52, this interval will have the population mean 95% of the time. In general, we calculate the 95% confidence interval for the mean with the following formula:

$$\text{Lower limit} = M - Z_{.95}\sigma_M$$

$$\text{Upper limit} = M + Z_{.95}\sigma_M$$

Where $Z_{.95}$ is the number of standard deviations extending from the mean of a normal distribution needed to contain 0.95 of the area and σ_M is the standard error of the mean.

If we look closely at this formula for a confidence interval, we will notice that we require to know the standard deviation (σ) in order to estimate the mean. This might sound unrealistic and it is. However, computing a confidence interval when σ is known is easier than when σ has to be estimated and serves a pedagogical purpose.

Suppose the following 5 numbers were sampled from a normal distribution with a standard deviation of 2.5: 2, 3, 5, 6 and 9. To compute the 95% confidence interval, start by computing the mean and standard error:

$$M = (2+3+5+6+9)/5 = 5$$

$$\sigma_M = 1.118$$

Z.95 can be found using the normal distribution calculator and specifying that the shaded area is 0.95 and indicating that we want the area to be between the cutoff points. As represented in the figure below, the value is 1.96. If we wanted to compute the 99% confidence interval, we would have set the shaded area to 0.99 and the result would have been 2.58.

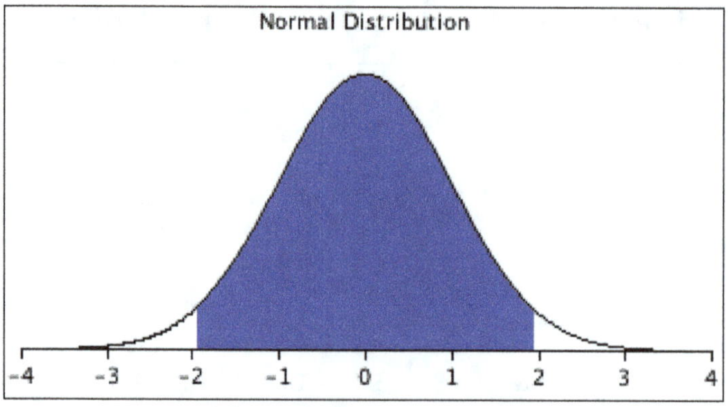

2. 95% of the area is between -1.96 and 1.96.

2.1.2 Statistical Hypothesis

A statistical hypothesis is an assumption about the population parameter. This assumption might or might not be true. Hypothesis testing defines the formal procedures used by statisticians to accept or reject statistical hypotheses.

The best way to find out whether a statistical hypothesis is true would be to investigate the entire population. Since that is often impractical, researchers typically examine a random sample from the population. If sample data are not consistent with the statistical hypothesis, the hypothesis is rejected.

There are 2 types of statistical hypotheses:

- Null hypothesis: The null hypothesis, referred as Ho, it is usually the hypothesis that sample observations result purely from chance.
- Alternative hypothesis: The alternative hypothesis, referred as H1 or Ha, it is the hypothesis that sample observations are influenced by some non-random cause.

For example, suppose we need to determine whether a coin was fair and balanced. A null hypothesis may be that half the flips would result in Heads and half, in Tails.

The alternative hypothesis may be that the number of Heads and Tails would be very different. Symbolically, these hypotheses would be termed as:

$$H_0 : P = 0.5$$

$$H_a : P \neq 0.5$$

If we flipped the coin 50 times, resulting in 40 Heads and 10 Tails. Given this result, we would be inclined to reject the null hypothesis. We would conclude, based on the evidence, that the coin was probably not fair and balanced.

2.1.3 Null and Alternative Hypothesis

Null Hypothesis

For applying the test of significance, we first set up a hypothesis which is a statement about the population parameter. This statement is usually a hypothesis of no difference and it is denoted by Ho.

Null Hypothesis H_0: Two samples have been drawn from the same population of (S.D) $\sigma = 2.5$ inches.

$$H_0 : \mu_1 = \mu_2 \ \& \ \sigma = 2.5 \text{ inches}$$

$$H_1 : \mu_1 \neq \mu_2.$$

The test statistic is $Z = \dfrac{\overline{x}_1 - \overline{x}_2}{\sqrt{\dfrac{\sigma^2}{n_1} + \dfrac{\sigma^2}{n_2}}}$

$$Z = \frac{67.5 - 68}{\sqrt{2.5^2\left(\dfrac{1}{1000} + \dfrac{1}{2000}\right)}} = -5.16$$

$$Z = 5.16.$$

$$\text{Here } Z = 5.16 > 1.96.$$

∴ The null hypothesis Ho is rejected at 5% level of significance.

i.e., The samples are not drawn from the same population of standard deviation S.D = 2.5 inches.

Alternative Hypothesis

The null hypothesis is false and then something else must be true. This is called an alternate hypothesis and it is denoted by H_1.

2.2 Level of Significance

Connection Between Type I Error and Significance Level

A significance level α corresponds to a certain value of the test statistic, refers tα, shown by the orange line in the picture of a sampling distribution below. The picture represents a hypothesis test with alternate hypothesis "μ > 0".

Since the shaded area referred by the arrow is the p-value corresponding to tα, that the p-value (shaded area) is α. To have the p-value less than α, a t-value for this test should be to the right of tα. So the probability of rejecting the null hypothesis when it is true is the probability that t > tα, which we saw the above is α. In other words, the probability of Type I error is α[1].

The significance level α is the probability of creating the wrong decision when the null hypothesis is true.

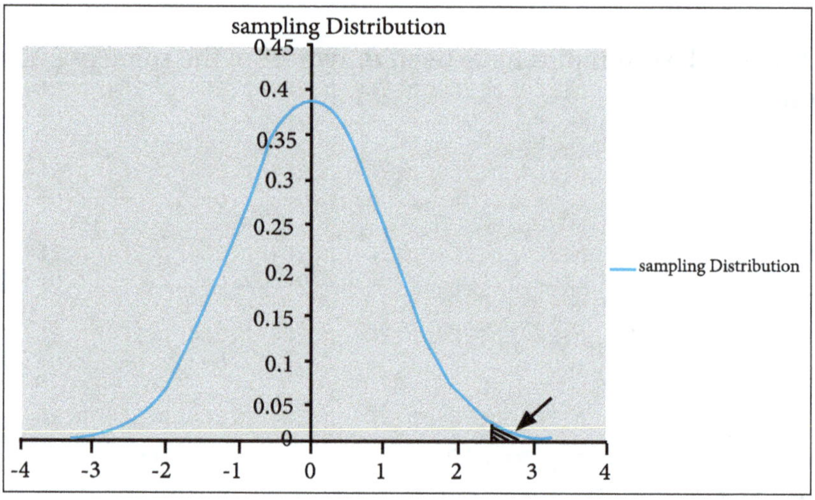

Pros and Cons of Setting a Significance Level

- Setting a significance level (before doing inference) has the advantage that the analyst is not tempted to chose a cut-off on the basis of what he or she hopes is true.
- It has the drawback that it neglects that some p-values may best be considered borderline. This is one reason. Why it is necessary to report p-values when reporting results of hypothesis tests. It is also good practice to include confidence intervals corresponding to the hypothesis test.

Type II error: Not rejecting the null hypothesis when in fact the alternate hypothesis is true is known as a Type II error.

Note: “The alternate hypothesis” in the definition of Type II error might refer to the alternate hypothesis in a hypothesis test or it refer to a “specific” alternate hypothesis.

Example: In a t-test for a sample mean μ, with null hypothesis “$\mu = 0$” and alternate hypothesis “$\mu > 0$”, we might talk about the Type II error relative to the general alternate hypothesis “$\mu > 0$” or may talk about the Type II error relative to the specific alternate hypothesis “$\mu > 1$”. Note that the specific alternate hypothesis is a special case of the common alternate hypothesis.

In practice, people often work with Type II error relative to a specific alternate hypothesis. In this situation, the probability of Type II error relative to the specific alternate hypothesis is often called β. In other words β is the probability of making the false decision when the specific alternate hypothesis is true.

Considering both types of error together: The below table explains Type I and Type II errors:

		Truth (for the population studied)	
		Null Hypothesis True	Null Hypothesis False
Decision (based on the sample)	Reject Null Hypothesis	Type I Error	Correct Decision
	Fail to reject Null Hypothesis	Correct Decision	Type II Error

The following diagram represents the Type I error and the Type II error against the particular alternate hypothesis “$\mu = 1$” in a hypothesis test for a population mean μ, with null hypothesis “$\mu = 0$,” alternate hypothesis “$\mu > 0$” and significance level $\alpha = 0.05$.

- The blue (leftmost) curve represented below is the sampling distribution assuming the null hypothesis “$\mu = 0$.”
- The green (rightmost) curve represented below is the sampling distribution, it assuming the specific alternate hypothesis “$\mu = 1$”.
- The vertical red line represented the cut-off for rejection of the null hypothesis:
    - The area of the diagonally hatched region to the right of the red line and under the blue curve is the probability of type I error (α).
    - The area of the horizontally hatched region to the left of the red line and under the green curve is the probability of Type II error (β).
    - The null hypothesis is ignored for the values of the test statistic to the right of the red line and not rejected for the values to the left of the red line.

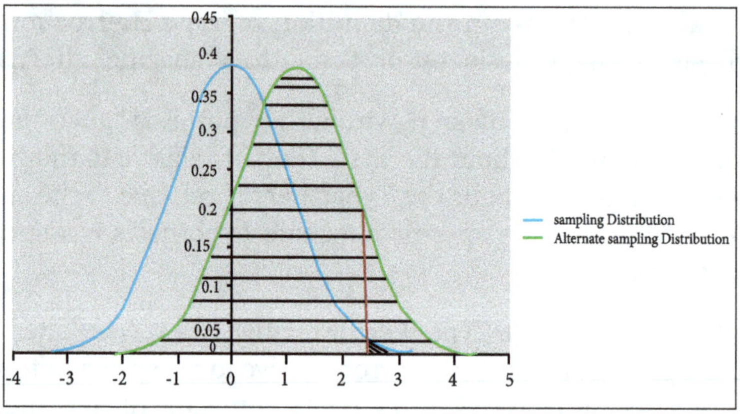

Deciding what significance level to use:

- This must be done before analyzing the data preferably before gathering the data. The choice of significance level should be based on the consequences of Type I and Type II errors.
- If the consequences of a type I error are serious or expensive, then a very small significance level is appropriate.

  Example 1: 2 drugs are being compared each other for effectiveness in treating the same condition. Drug 1 is very affordable, but Drug 2 is extremely expensive. The null hypothesis can "both drugs are equally effective" and the alternate is "Drug 2 is more effective than Drug 1." In such situation, a Type I error would be deciding that Drug 2 is more effective, when in fact it is no better than Drug 1, but would cost the patient much more money. That will be undesirable from the patient's perspective so a small significance level is warranted.

Problems

1. The heights of college students in Chennai are normally distributed with standard deviation 6 cm and sample of 100 students had their mean height 158 cm. Let us test the hypothesis that the mean height of college students in Chennai is 160 cm at 1 % level of significance.

Solution:

Given data:

$n = 100; \bar{x} = 158$

$n = 160$

$\sigma = 6$

$H_0 : \bar{x} \neq \mu$ (The difference between x and μ is not significant).

$H_1 : \bar{x}\ \mu$

$$z = \frac{\bar{x}-\mu}{\sigma/\sqrt{n}} = \frac{158-160}{6/10} = -3.33$$

$|Z| = 3.33$

WKT $Z_{0.01} = 2.58$; Here $|Z| > Z_{0.01}$

$\therefore$ H_o is rejected

$\therefore$ The sample is not drawn from the population with mean height $\mu = 160$ cm.

2. A machine produces 16 imperfect articles in a sample of 500. After machine is over hauled, it produces 3 imperfect articles in a batch of 100. Let us check if the machine has been improved.

Solution:

Given data:

$n_1 = 500$

$n_2 = 100$

$$p_1 = \frac{16}{500} = 0.032$$

$$p_2 = \frac{3}{100} = 0.03$$

$H_o : P_1 = P_2 = P.$

$H_1 : P_1 \neq P_2$

The test statistic is:

$$Z = \frac{p_1 - p_2}{\sqrt{pq\left(\frac{1}{n_1} + \frac{1}{n_2}\right)}}$$

Where $p = \frac{n_1 p_1 + n_2 p_2}{n_1 + n_2}$

$$= \frac{500(0.032) + 100(0.03)}{500 + 100} = \frac{19}{600}$$

$= 0.0316$

$$q = 1 - 0.0316$$

$$= 0.9684$$

$$\therefore Z = \frac{0.032 - 0.03}{\sqrt{(0.0316)(0.9684)\left(\frac{1}{500} + \frac{1}{100}\right)}}$$

$$= \frac{0.002}{\sqrt{(0.0307)(0.012)}} = \frac{0.002}{0.0191}$$

$$|Z| = 0.104$$

Since, | Z |< 1.96. We accept the null hypothesis at level of significance.

3. A manufacturer of bulbs claims that an average of 2% of the bulbs manufactured by him is defective. A random sample of 400 bulbs contained 13 defective bulbs. On the basis of the sample, let us check if we can support the manufacturers claim at 5% level of significance.

Solution:

Given data:

$$n = 400$$

Sample proportion $p = \frac{13}{400} = 0.0325$

Population proportion $P = 2\% = 0.02.$

$$\therefore Q = 1 - P = 1 - 0.02 = 0.98.$$

Null Hypothesis $H_0 : P = 0.02$

Alternate Hypothesis H_1: P > 0.02 (one tail test)

The test statistic $z = \frac{p - P}{\sqrt{\frac{PQ}{n}}}$

$$z = \frac{0.0325 - 0.02}{\sqrt{\frac{(0.02)(0.98)}{400}}} = 1.7857$$

$$|z| = 1.7857$$

At 5% level of significance for one tail test $z_\alpha = 1.645$.

Here calculated z value = 1.7857 is greater than tabulated z_α Value = 1.645.

Hence we reject the null hypothesis.

2.2.1 Test of Significance

Test of Significance for Large Samples

For large n (sample size), almost all the distributions can be approximated very closely by a normal probability curve, we therefore use the normal test of significance for large samples:

$$Z = \frac{t - E(t)}{\sqrt{V(t)}} = N(0.1)$$

Thus if the discrepancy between the observed and the expected (hypothetical) value of a statistics is greater than Zα times the standard error (S.E), hypothesis is rejected at α level of significance. Similarly to give by:

$$t - E(t) \le Z\alpha \text{ S.E } (t),$$

The deviation is not regarded significant at 5% level of significance. In other words the deviation t- E(t), could have arisen due to fluctuations of sampling and the data do not provide any evidence against the null hypothesis which may, therefore be accepted at α level of significance.

If $Z \le 1.96$, then the hypothesis Ho is accepted at 5% level of significance.

Thus the steps to be used in the normal test are as follows:

- Compute the test statistic Z under Ho.
- If $Z > 3$, Ho is always rejected.
- If $Z < 3$, we test its significance at certain level of significance.

2.2.2 Test Based on Normal Distribution

There are two types of samples in Testing of Hypothesis:

Small Sample: (n <30)

- Type 1 Student's 't' test.
- Single Mean (S.D given directly).

- Single Mean (S.D is not given directly).
- Difference of Means.

Large Sample: (n >30)

- Type 1 Single Mean.
- Type 2 Single Proportion.
- Type 3 Difference of Mean.
- Type 4 Difference of Proportion.

2.2.3 Test for Means and Proportions

Test for Single Mean

The tests of significance for variables are that the sample mean is asymptotically normally distributed even if the parent population from which the sample is drawn is not normal.

If xi (i =1 ,n) is a random sample of size n from a normal population with mean and variance σ2, then the sample mean is distributed normally with mean and variance $\frac{\sigma^2}{n}$.

$$\bar{x} \sim N\left(\mu, \frac{\sigma^2}{n}\right)$$

H_0: There is no significant difference between sample mean $(\bar{x})$ and population mean.

Test Statistic:

$$Z = \frac{\bar{x} - \mu}{\sigma/\sqrt{n}} \sim N(0,1)$$

If σ^2 is unknown, then it is estimated by sample variance i.e., $\sigma^2 = s^2$ (for large n).

Test for Difference of Means

Let $\bar{x}_1 (\bar{x}_2)$ be the mean of a sample of size n_1 (n_2) from a population with mean μ1 (μ2) and variance σ_1^2 (σ_2^2). Therefore:

$$\bar{x}_1 \sim N\left(\mu_1, \frac{\sigma_1^2}{n_1}\right), \ \bar{x}_2 \sim N\left(\mu_2, \frac{\sigma_2^2}{n_2}\right)$$

The difference $(\bar{x}_1 - \bar{x}_2)$ is also normal variate test statistic:

$$Z = \frac{\bar{x}_1 - \bar{x}_2 - E(\bar{x}_1 - \bar{x}_2)}{S.E\ (\bar{x}_1 - \bar{x}_2)} \sim N(0,1)$$

$$Z = \frac{\bar{x}_1 - \bar{x}_2 - (\mu_1 - \mu_2)}{\sqrt{\frac{\sigma_1^2}{n_1} + \frac{\sigma_2^2}{n_2}}}$$

Under the null hypothesis $= H_o : \mu_1 = \mu_2$

$$Z = \frac{(\bar{x}_1 - \bar{x}_2)}{\sqrt{\frac{\sigma_1^2}{n_1} + \frac{\sigma_2^2}{n_2}}} \sim N(0,1)$$

Therefore, $Z = \frac{(\bar{x}_1 - \bar{x}_2)}{\sigma\sqrt{\frac{1}{n_1} + \frac{1}{n_2}}}$, if $\sigma_1^2 = \sigma_2^2 = \sigma^2$

If σ is not known, then its estimate is used $\hat{\sigma}^2 = \frac{n_1 s_1^2 + n_2 s_2^2}{n_1 + n_2}$

Hypothesis testing applications with a continuous outcome variable in a single population are performed according to the five-step procedure outlined above. A key component is setting up the null and research hypotheses. The objective is to compare the mean in a single population to known mean (μo). The known value is generally derived from another study or report, for example a study in a similar, but not identical, population or a study performed some years ago. The latter is known as historical control. It is necessary in setting up the hypotheses in a one sample test which the mean specified in the null hypothesis is a fair and reasonable comparator.

In one sample tests for a continuous outcome, we set up our hypotheses against an appropriate comparator. We select a sample and compute descriptive statistics on the sample data - including the sample size (n), the sample mean $(\bar{X})$ and the sample standard deviation (s). We then determine the appropriate test statistic (Step 2) for the hypothesis test. The formulas for test statistics depend on the sample size and are given below.

H_o: $\mu = \mu_o$

Test Statistics for Testing $H_o : \mu = \mu_o$	
if $n \geq 30$	if $n < 30$
$Z = \frac{\bar{X} - \mu_o}{s/\sqrt{n}}$	$t = \frac{\bar{X} - \mu_o}{s/\sqrt{n}}$ where df=n-1

2.3 Small Samples

The approach is appropriate as long as the sample includes at least one success and one failure. The key steps are:

- Formulate the hypotheses to be tested: This means stating the null hypothesis and the alternative hypothesis.
- Determine the sampling distribution of the proportion: If the sample proportion is the result of a binomial experiment, the sampling distribution will be binomial. If it is the outcome of a hyper geometric experiment, the sampling distribution will be hyper geometric.
- Specify the significance level: (Researchers often set the significance level equal to 0.05 or 0.01, although other values might be used.)
- Based on the hypotheses, the sampling distribution and the significance level, define the section of acceptance.
- Test the null hypothesis: If the sample proportion falls within the region of acceptance, accept the null hypothesis; otherwise, reject the null hypothesis.

Several surveys deal with large populations, such as all adults in Canada or all adults in Saskatchewan. The target populations of these surveys can has hundreds of 1000's or even millions of people. The sampling and statistical analysis process assume that:

- The population is very large and the size of the sample is small when compared to the size of the population.
- When the target population is less than approximately 5000 or if the sample size is a important proportion of the population size, such as 20% or more, then the standard sampling and statistical analysis techniques require to be changed.

Typical situations involving small populations include:

- Sampling residents of a small town or small city.
- Sampling employees in a firm.
- Sampling members of a trade or professional group.
- Sampling cows from a herd of cattle.
- Sampling hospitals from around the country.
- Sampling emissions from power plants in Canada.

In each of these cases the target population may vary from several dozen to several thousand.

When dealing with large populations, the sample size is determined using the normal approximation to the binomial distribution. This approximation is very accurate when the population is large and the sample size is small. However, if we were to sample a population of 200 individuals, then for a given accuracy, we would require a far smaller sample than that calculated using the normal approximation to the binomial. To determine the sample size for small populations, we utilize the normal approximation to the hyper geometric distribution. The sample size formulas for large (binomial) and small (hypergeometric) populations are represented below.

$$\text{binomial}: \quad n = \frac{z^2 pq}{E^2}$$

$$\text{hypergeometric}: \quad n = \frac{Nz^2 pq}{\left(E^2 (N-1) + z^2 pq\right)}$$

Where,

- n is the needed sample size.
- N is the population size.
- p and q are the population proportions. (If we don't know what these, are set them each to 0.5.
- z is the value that specifies the level of confidence we need in our confidence interval when we analyze the data. The typical levels of confidence for surveys are 95%, in which case z is set to 1.96.
- E sets the accuracy of the sample proportions. If we want to know what proportion of individuals are in favour of some policy, with an accuracy of plus or minus 3%, then E is set to 0.03.

For very small populations (50 or less), we need almost the entire population in order to achieve accuracy.

There is a limit on the accuracy we can achieve when dealing with small populations. The maximum resolution we can achieve for a proportion is 1/N. If E is set smaller than this, we can never reach this accuracy even if we survey the entire population.

If we are carrying out an opinion survey from a large population and we want the confidence intervals for our sample proportions to have a 95% confidence level with an error of plus or minus 0.03%, then the minimum sample size required is 1,068 individuals.

If we were to sample a small population, the sample size would be much smaller:

Population Size	Required Sample Size
5000	880
1000	517
500	341
300	235

As the population size becomes smaller than 300, we might as well survey everyone in the population.

Problem 1: Sampling with Replacement

If an urn has 30 marbles. Some marbles are red and the rest are green. A researcher hypothesizes that the urn contains 15 or more red marbles. The researcher randomly samples 5 marbles, with replacement, from the urn. 2 of the selected marbles are red and three are green. Based on the sample results, should the researcher accept or reject the hypothesis. Use a significance level of 0.20.

Solution:

There are 5 steps in conducting a hypothesis test, as described above. We work through each of the five steps below:

Formulate hypotheses: The initial step is to state the null hypothesis and an alternative hypothesis.

Null hypothesis: P >= 0.50

Alternative hypothesis: P < 0.50

Note that these hypotheses constitute a one-tailed test. The null hypothesis will be rejected only if the sample proportion is too small.

Find out sampling distribution: Since we sampled with replacement, the sample proportion can be considered an outcome of a binomial experiment. And based on the null hypothesis, we assume that at least 15 of 30 marbles are red. Thus, the true population proportion is assumed to be 15/30 or 0.50.

Provided those inputs (a binomial distribution where the true population proportion is equal to 0.50), the sampling distribution of the proportion can be found out. It appears in the table below.

Number of red marbles in sample	Sample proportion	Binomial probability	Cumulative probability
0	0.0	0.03125	0.03125
1	0.2	0.15625	0.1875
2	0.4	0.3125	0.5
3	0.6	0.3125	0.8125
4	0.8	0.15625	0.96875
5	1.0	0.03125	1.00

Specify the significance level: The significance level was the set at 0.20. (This means that the probability of occurring a Type I error is 0.20, assuming that the null hypothesis is true.)

The region of acceptance: From the sampling distribution (refer above table), we see that it is not possible to define a region of acceptance for which the significance level is exactly 0.20.

However, we can define the region of acceptance for which the significance level would be not more than 0.20. From the table, we see that if the true population proportion is equal to 0.50, we would be very unlikely to pick 0 or 1 red marble in our sample of 5 marbles. The probability of selecting 1 or 0 red marbles would be 0.1875. Therefore, if we let the significance level equal 0.1875, we can define the region of rejection as any sampled result that includes only 0 or 1 red marble (i.e., a sampled proportion equal to 0 or 0.20). We can define the region of acceptance as any sampled outcome that includes at least two red marbles. This is equivalent to a sampled proportion that is greater than or equal to 0.40.

Test the null hypothesis: Since the sample proportion (0.40) is in the region of acceptance, we can't reject the null hypothesis.

Problem 2: Sampling without Replacement

The Acme Advertising Company has twenty five clients. Account executives at Acme claim that 80% of these clients are very satisfied with the service they receive. To test that claim, Acme's CEO commissions a survey of 10 clients. Survey participants are randomly sampled, without a replacement, from the client population. Six of the ten sampled customers (i.e., 60%) say that they are very satisfied. Based on the sample results, should the CEO accept or reject the hypothesis that 80 percent of Acme's clients are very satisfied. Use a significance level of 0.10.

Solution:

There are five steps in conducting a hypothesis test, as described above. We work via each of the 5 steps below:

1. Formulate hypotheses: The initial step is to state the null hypothesis and an alternative hypothesis.

- Null hypothesis: P >= 0.80
- Alternative hypothesis: P < 0.80

Note that these hypotheses constitute a one-tailed test. The null hypothesis will be eliminated only if the sample proportion is too small.

2. Find out sampling distribution: Since we sampled without the replacement and the sample proportion can be considered an outcome of a hypergeometric experiment. Based on the null hypothesis, we can assume that at least 80 percent of the 25 clients (that is 20 clients) are very satisfied.

Provide those inputs (a hypergeometric distribution where 20 of 25 clients are very satisfied), the sampling distribution of the proportion can be found out. It appears in the table below.

Number of satisfied clients in sample	Sample proportion	Hypergeometric probability	Cumulative probability
4 or less	0.4 or less	0.00	0.00
5	0.5	0.00474	0.00474
6	0.6	0.05929	0.06403
7	0.7	0.23715	0.30119
8	0.8	0.38538	0.68656
9	0.9	0.25692	0.94348
10	1.0	0.05652	1.00

3. Specify the significance level: The significance level was set at 0.10. It represents that the probability of making a Type I error is 0.10, considering that the null hypothesis is true.

4. Region of acceptance: From the sampling distribution (refer the above table), we can see that it is not possible to define a region of acceptance for which the significance level is exactly 0.10.

However, we can define a region of acceptance for which the significance level would be no more than 0.10. From the table, we can see that if the exact proportion of very satisfied clients is equal to 0.80, we would be very dissimilar to have fewer than 7 very satisfied clients in our sample. The probability of having six or fewer very satisfied clients in the sample will be 0.064. Therefore, if we let the significance level equal 0.064, we can define the region of rejection as any sampled outcome that includes 6 or fewer very satisfied customers. We can describe the region of acceptance as any sampled outcome that includes 7 or more very satisfied customers. This is equivalent to a sample proportion that is greater than or equal to 0.70.

5. Test the null hypothesis: Since the sample proportion (0.60) is outside the region of acceptance, we can't accept the null hypothesis at the 0.064 level of significance.

2.3.1 T- Test for One Sample and Two Sample Problem and Paired T-Test

Tests Based on T-Distribution

1. Test for an Assumed Population Mean:

A random sample $x_1,..,x_n$ of size n ($n \geq 2$) has been drawn from a normal population whose variance σ^2 is unknown. On the basis of this random sample the aim is to test the following:

$$H_o : \mu = \mu_o$$

$$H_o : \mu \neq \mu_o \ (\text{two-tailed})$$

$$\mu > \mu_o \ (\text{right-tailed})$$

$$\mu < \mu_o \ (\text{left-tailed})$$

Test statistic:

$$t = \frac{\bar{x} - \mu_o}{s/\sqrt{n}} \sim t_{n-1},$$

$$\text{where } \bar{x} = \frac{1}{n}\sum_{i=1}^{n} x_i \text{ and } s^2 = \frac{1}{n-1}\sum_{n-1}^{n}(x_i - \bar{x})^2$$

The table giving the value of t required for significance at various levels of probability and for different degrees of freedom is called the t–tables. The computed value is compared with the tabulated value at 5 or 1 percent levels of significance and at (n-1) degrees of freedom and accordingly the null hypothesis is accepted or rejected.

2. Test for the Difference of Two Population Means:

Let $x_1(x_2)$ be the sample mean of a sample of size n_1 (n_2) from a population with mean 1 (2) and variance of the two population be same σ2, which is unknown.

$$H_o : 1 - 2 = \delta_o$$

$$\text{Since } \bar{x}_1 \sim N(\mu_1, \sigma^2/n_1)$$

$$\bar{x}_2 \sim N(\mu_2, \sigma^2/n_2)$$

$$\bar{x}_1 - \bar{x}_2 \sim N\left(\mu_1 - \mu_2, \frac{\sigma^2}{n_1} + \frac{\sigma^2}{n^2}\right)$$

Test statistic: Under H_o,

$$t = \frac{\bar{x}_1 - \bar{x}_2 - \delta_o}{\sigma\sqrt{\frac{1}{n_1} + \frac{1}{n_2}}} \sim t_{n_1+n_2-2}$$

Since σ^2 is unknown, therefore, it is estimated from the sample,

$$s^2 = \frac{(n_1 - 1)s_1^2 + (n_2 - 1)s_2^2}{n_1 + n_2 - 2}$$

$$t = \frac{\bar{x}_1 - \bar{x}_2 - \delta_o}{s\sqrt{\frac{1}{n_1} + \frac{1}{n_2}}} \sim t_{n_1+n_2-2}$$

If $\delta_o = 0$, this test reduces to test the equality of two population means.

The R function t. test () can be employed to perform both one and 2 sample t-tests on vectors of data.

The function has a variety of options and can be called as follows:

t. test(x, y = NULL, alternative = c("two. sided", "less", "greater"), mu = 0, paired = FALSE, var. equal = FALSE, conf. level = 0.95)

- Here x is a numeric vector of the data values and y is an optional numeric vector of the data values.
- If y is excluded, the function performs a one-sample t-test on the data contained in x, if it is included it performs a 2-sample t-tests using both x and y.

- The option mu gives a number indicating the true value of the mean (or difference in means if we are performing a 2 sample test) under the null hypothesis.
- The option alternative is a character string specifying the alternative hypothesis and must be one of the following: "two. sided" (which is the default), "greater" or "less" depending on whether the alternative hypothesis is that the mean is different than, greater than or less than mu, respectively.

For example,

The following call:

t. test(x, alternative = "less", mu = 10)

performs one sample t-test on the data contained in x where the null hypothesis is that =10 and the alternative is that <10.

The option paired denotes whether or not we want a paired t-test (TRUE = yes and FALSE = no). If we leave this option out it defaults to FALSE.

The option var. equal is a logical variable indicating whether or not to assume the 2 variances as being equal when performing a 2-sample t-test. If TRUE then the pooled variance is used to estimate the variance otherwise the Welch (or Satterthwaite) approximation to the degrees of freedom is used. If we leave this option out it defaults to FALSE.

At last, the option conf. level find out the confidence level of the reported confidence interval for in the one-sample case and 1- 2 in the two-sample case.

One-Sample T-Tests

Example:

Let us consider, an outbreak of Salmonella-related illness was attributed to ice cream produced at an certain factory. Scientists measured the level of Salmonella in 9 randomly sampled batches of ice cream. The levels (in MPN/g) were:

0.593,0.142, 0.329, 0.691, 0.231, 0.793, 0.519, 0.392, 0.418

Is there evidence that the mean level of Salmonella in the ice cream is greater than 0.3 MPN/g?

Let be the mean level of Salmonella in all batches of ice cream. Here the hypothesis of interest can be termed as:

$$H_o := 0.3$$

$$H_a : > 0.3$$

Hence, we will require to include the options alternative="greater", mu=0.3. Below is the relevant R-code:

- x = c(0.593, 0.142, 0.329, 0.691, 0.231, 0.793, 0.519, 0.392, 0.418)
- t. test(x, alternative="greater", mu=0.3)

One Sample t-test:

data: x

t = 2.2051, df = 8, p-value = 0.02927

alternative hypothesis: true mean is greater than 0.3

From the output we can see that the p-value = 0.029. Hence, there is moderately strong evidence that the mean Salmonella level in the ice cream is above 0.3 MPN/g.

Two-Sample T-Tests

Example:

Six subjects were provided a drug (treatment group) and an additional 6 subjects a placebo (control group). Their reaction time to a stimulus was measured (in ms). We need to perform a two-sample t-test for comparing the means of the treatment and control groups.

Let 1 be the mean of the population taking medicine and 2 the mean of the untreated population. Here the hypothesis of interest can be termed as:

$$H_o : 1 - 2 = 0$$

$$H_a : 1 - 2 < 0$$

Here we will require including the data for the treatment group in x and the data for the control group in y. We will also require to include the options alternative="-less", mu=0.

Finally, we need to decide whether or not the standard deviations are the same in both groups.

Below is the relevant R-code when assuming equal standard deviation:

- Control = c (91, 87, 99, 77, 88, 91)
- Treat = c (101, 110, 103, 93, 99, 104)
- t. test (Control, Treat, alternative="less", var. equal=TRUE)

Two-Sample T-Test:

Data: Control and Treat.

t = -3.4456, df = 10, p-value = 0.003136

Alternative hypothesis: true difference in means is less than 0.

Below is the relevant R-code when not assuming equal standard deviation:

- t. test (Control, Treat, alternative="less")

Welch 2 Sample t-test:

Data: Control and Treat.

t = -3.4456, df = 9.48, p-value = 0.003391

Alternative hypothesis: true difference in means is less than 0.

Here the pooled t-test and the Welsh t-test give roughly the same results (p-value = 0.00313 and 0.00339, respectively).

Paired T-Tests

There are several experimental settings where each subject in the study is in both the treatment and control group. For example, in a matched pairs design, subjects are matched in pairs and different treatments are provided to each subject in the pair. The result are thereafter compared pair-wise. Alternatively, one can measure each subject twice, before and after a treatment. In either of these situations we cannot use 2-sample t-tests since the independence assumption is not valid. Instead we need to use a paired t-test. This can be done using the option paired =TRUE.

2.3.2 F-Test and Chi-Square Test (Testing of Goodness of Fit and Independence)

Test of Significance based on Chi-Square Distribution

Test for the Variance of a Normal Population:

Let $x_1, x_2, \ldots, x_n \ (n \geq 2)$ be a random sample from $N(\mu, \sigma^2)$

$H_0 : \sigma^2 = \sigma_0^2$

Test statistic :

$$\chi^2 = \sum_{i=1}^{n} \left(\frac{x_i - \mu}{\sigma_0} \right)^2 \sim \chi_n^2, \text{ when } \mu \text{ is known,}$$

$$\chi^2 = \sum_{i=1}^{n}\left(\frac{x_i - \bar{x}}{\sigma_o}\right)^2 \sim \chi^2_{n-1}, \text{ when } \mu \text{ is not known,}$$

Tables are available for χ^2 at different levels of significance and with different degrees of freedom.

Applications of CHI Square Test

- The sample observations should be independent.
- Constraints on the cell frequencies, if any, must be linear.

Test of Significance Based on F-Distribution

Test for the Comparison of Two Population Variances:

Let $x_i, i = 1, \ldots, n_1$ and $x_j, j=1, \ldots, n_2$ be the two random sample of sizes n_1 and n_2 drawn from two independent normal populations $N(\mu_1, \sigma_1^2)$ and $N(\mu_1, \sigma_2^2)$ respectively.

S_1^2 and S_2^2 are the sample variances of the two samples:

$$S_1^2 = \frac{1}{n_1 - 1}\sum_{i-1}^{n_1}(x_i - \bar{x}_1)^2 \text{ and } S_2^2 = \frac{1}{n_2 - 1}\sum_{j-1}^{n_2}(x_j - \bar{x}_2)^2$$

$$\bar{x}_1 = \frac{1}{n_1}\sum_{i-1}^{n_1} x_i, \quad \bar{x}_2 = \frac{1}{n_2}\sum_{j-1}^{n_2} x_j$$

H_o: The ratio of two population's variances is specified:

$$\frac{\sigma_1^2}{\sigma_2^2} = \sigma_o^2$$

Test statistic: Assuming $S_1^2 > S_2^2$,

$$F = \frac{\sigma_2^2}{\sigma_1^2}\frac{S_1^2}{S_2^2} \sim F_{n1-1,\, n2-1} \text{ under } H_o, F = \frac{S_1^2}{S_2^2}\frac{1}{\sigma_2^2} \sim F_{n1-1,\, n2-1}$$

Problems

1. Let us examine whether the difference in the variability in yields is significant at 5% level of significance, for the following.

	Set of 40 plots	Set of plots
Mean yield per plot	1256	1243
S.D. per plot	34	28

Solution:

H_0: There is no significant difference in the variability in yields.

Since two sample means $\overline{x}_1$ and $\overline{x}_2$ are given and also sample standard deviation S1 and S2 are given we use the statistic.

$$Z = \frac{\overline{x}_1 - \overline{x}_2}{\sqrt{S^2\left(\frac{1}{n_1} + \frac{1}{n_1}\right)}}$$

Where,

$$S^2 = \frac{1}{n_1 + n_2 - 2}\left[n_1 s_1^2 + n_2 s_2^2\right]$$

$$= \frac{1}{40+60-2}\left[40(34)^2 + 60(28)^2\right]$$

$$= \frac{1}{98}\left[40(34)^2 + 60(28)^2\right]$$

$$= 951.837$$

$$\therefore Z = \frac{1258 - 1243}{\sqrt{951.837\left(\frac{1}{40} + \frac{1}{60}\right)}}$$

$$Z = 2.382$$

$$|Z| = 1.96$$

$\therefore$ H_0 is rejected at 5% level of significance. There is a significant difference in the variability in yields.

2. A coin is tossed 400 times and it turns up head 216 times. Let us discuss whether the coin may be unbiased one at 5% level of significance.

Solution:

Given:

$$n = 400$$

P = Probability to getting a head in a toss $= \frac{1}{2}$

$$Q = 1 - P = \frac{1}{2}$$

X = Number of successes = 2.16

H_0: The coin is unbiased.

H_1: The coin is biased.

$$Z = \frac{X - np}{\sqrt{nPQ}} = \frac{216 - (400)\left(\frac{1}{2}\right)}{\sqrt{400(1/2)(1/2)}} = 1.6$$

$$Z_{0.05} = 1.96$$

$$\text{Here } Z(=1.6) < Z_{0.05}(=1.96)$$

Hence, we accept H_0 at 5% LOS i.e., the coin is unbiased.

3. A Out of 8000 graduates in a town 800 are females, out of 1600 graduate employees 120 are females. Let us use x^2 to determine if any distinction is made in appointment on the basis of sex. Value of x^2 at 5% level for one degree of freedom is 3.84.

The given information can be given in a tabular form:

	Employment	Unemployment	Total
Males	1480	5720	7200
Females	120	680	800
total	1600	6400	8000

Let us take the hypothesis that no distinction is made in appointment on the basis of sex. Applying x^2 test.

Solution:

$$E_{11} = \frac{7200}{8000} \times 1600$$

$$= 1440$$

The table of expected frequencies is given below:

	1440	5760	7200
	160	640	800
total	1600	6400	8000
o	E	$(o-E)^2$	$(o-E)^2/E$
1480	1440	1600	1.111
120	160	1600	10.000
5720	5760	1600	0.278
680	640	1600	2.500
			13.889

$\therefore X^2 = 13.889$

$v = (r-1)(-1) = (2-1) = 1$

for $v = 1$, $X_{0.03} = 3.84$

4. Let us test if the difference in the mean is significant for the following data.

Sample I:	76	68	70	43	94	68	33	
Sample II:	40	48	92	85	70	76	68	22

Solution:

Calculation for sample means and S.D's:

x	$x-\bar{x}$	$(x-\bar{x})^2$	y	$y-\bar{y}$	$(y-\bar{y})^2$
76	11.4	129.96	40	−22.6	510.76
68	3.4	11.56	48	− 14.6,	213.16
70	5.4	29.16	92	29.4	864.36
43	−21.6	466.56	85	22.4	501.76
94	29.4	864.36	70	7.4	54.76
68	3.4	11.56	76	13.4	179.56
33	−31.6	998.56	68	5.4	29.16
			22	−40.6	1648.36
452		2511.72	501		4001.88

$$\bar{x} = \frac{452}{7} = 64.5 \quad \sum(x-\bar{x})^2 = 2511.72$$

$$\bar{y} = \frac{501}{8} = 62.5 \quad \sum(y-\bar{y})^2 = 4001.88$$

$$\therefore S^2 = \frac{1}{n_1 + n_2 - 2}\left[\sum(x-\bar{x})^2 + \sum(y-\bar{y})^2\right]$$

$$= \frac{1}{7+8-2}(2511.72 + 4001.88)$$

$$= 501.04$$

$$S = 22.38$$

H_0: There is no significant difference between the two means the test statistic is:

$$t = \frac{\bar{x}-\bar{y}}{\sqrt{S\frac{1}{n_1}+\frac{1}{n_2}}}$$

$$= \frac{64.6-62.6}{22.38\sqrt{\frac{1}{7}+\frac{1}{8}}}$$

$$= \frac{2}{22.38\sqrt{\frac{15}{56}}}$$

$$t = 0.1727$$

Tabulated value of 7 for 13 differences at 5% level of significance is 2.16.

Since calculation value < table value

$\therefore H_0$ is accepted.

i.e., there is no significant difference between the two means.

5. The following data gives the number of aircraft accidents that occurred during the various days of a week. Let us find whether the accidents are uniformly distributed over the week.

Days:	Sun	Mon	Tue	Wed	Thurs	Fri	Sat
No. of accidents:	14	16	8	12	1	9	14

Solution:

H_0: The accidents are uniformly distributed over the week.

Under this null hypothesis, the expected frequencies of the accidents on each of the days are $\frac{84}{7} = 12$.

O	E	$(O-E)^2$	$(o-E)^2/E$
14	12	4	0.333
16	12	16	1.333
8	12	16	1.333
12	12	0	0
11	12	1	0.0833
9	12	9	0.75
14	12	4	0.33
			4.1653

$$\psi^2 = \sum \frac{(O-E)^2}{E}$$

$$= 4.1653$$

difference $= n - 1$

$= 7 - 1$

$= 6$

$t0.05\ 6 = 12.592$

i.e., $\psi_2 = 11.07$

Calculation value < table value.

$\therefore$ H_0 is accepted,

i.e., the accidents are uniformly distributed over the week.

6. The means of two large samples of 1000 and 2000 members are 67.5 inches and 68.0 inches respectively. Let us check if the samples can be regarded as drawn from the same population of standard deviation 2.5 inches.

Solution:

Given:

$$n_1 = 1000; \; n_2 = 2000$$

$$\bar{x}_1 = 67.5 \text{ inches}; \; \bar{x}_2 = 68 \text{ inches}$$

Population standard deviation:

σ = 2.5 inches

Null Hypothesis H_0: Two samples have been drawn from the same population of (S.D) σ = 2.5 inches.

$$H_0 : \mu_1 = \mu_2 \;\; \& \;\; \sigma = 2.5 \text{ inches}$$

$$H_1 : \mu_1 \neq \mu_2$$

$$\text{The test statistic is } Z = \frac{\bar{x}_1 - \bar{x}_2}{\sqrt{\frac{\sigma^2}{n_1} + \frac{\sigma^2}{n_2}}}$$

$$Z = \frac{65.5 - 68}{\sqrt{(2.5)^2\left(\frac{1}{1000} + \frac{1}{2000}\right)}} = -5.16$$

$$|Z| = 5.16$$

$$\text{Here } |Z| = 5.16 > 1.96$$

∴ The null hypothesis Ho is rejected at 5% level of significance.

i.e., the samples are not drawn from the same population of standard deviation S.D = 2.5 inches.

7. Two independent samples of sizes 9 and 7 from a normal population had the following values of the variables.

Sample I:	18	13	12	15	12	14	16	14	15
Sample II:	16	19	13	16	18	13	15		

Let us check if the estimates of population variance differ significantly at 5% level of significance

Solution:

$$\text{Given: } n_1 = 9; \; n_2 = 7$$

Sample I X	X^2	Sample II Y	Y^2
18	324	16	256
13	169	19	361
12	144	13	169
15	225	16	256
12	144	18	324
14	196	13	169
16	256	15	225
14	196		
15	225		
Total $\sum X = 129$	$\sum X^2 = 1879$	$\sum Y = 110$	$\sum_y^2 = 1760$

$$\bar{x} = \frac{\sum x}{n_1} = \frac{129}{9} = 14.3333$$

$$\bar{y} = \frac{\sum y}{n_2} = \frac{110}{7} = 15.7143$$

$$s_1^2 = \frac{\sum x^2}{n_1} - \left(\frac{\sum y}{n_1}\right)^2 = \frac{1879}{9} - \left(\frac{129}{9}\right)^2$$

$$= 3.3342$$

$$s_2^2 = \frac{\sum y^2}{n_2} - \left(\frac{\sum y}{n_2}\right)^2 = \frac{1760}{7} - \left(\frac{110}{7}\right)^2$$

$$= 4.4894$$

$H_0 : \sigma_1^2 = \sigma_1^2$ (The difference of variance is not significant).

$H_1 : \sigma_1^2 \neq \sigma_2^2$

Degree of freedom:

$$v_1 = n_1 - 1 = 8;\ v_2 = n_2 - 1 = 6$$

The test statistics:

$$F = \frac{S_1^2}{S_2^2} \text{ or } \frac{S_2^2}{S_1^2}$$

$$S_1^2=\frac{n_1 s_1^2}{n_1-1}=\frac{9\times3.3342}{8}=3.751$$

$$S_2^2=\frac{n_2 s_2^2}{n_2-1}=\frac{7\times4.4894}{6}=5.2376$$

$$F=\frac{S_2^2}{S_1^2}=\frac{5.2376}{3.751}\left(\text{Here } S_2^2>S_1^2\right)$$

$$= 1.3963$$

Tabulated F value of 5% level of significance F5% (6, 8) = 3.58.

Calculated F = 1.3963 is less than Tabulated F value at 5% LOS.

∴ Hence Ho is accepted, i.e., there is no significance difference between population variance.

8. A sample of 10 boys had the I.Q's: 70, 120, 110, 101, 88, 83, 95, 98, 100 and 107. Let us test whether the population means I.Q may be 100.

Solution:

Given: n = 10; μ = 100

$$\bar{x}=\frac{\sum x}{n}=\frac{972}{10}=97.2$$

x	$x-\bar{x}$	$(x-\bar{x})^2$
70	−27.2	739.84
120	22.8	519.84
110	12.8	163.84
101	3.8	14.44
88	−9.2	84.64
83	−14.2	201.64
95	−2.2	4.84
98	0.8	0.64
100	2.8	7.84
107	9.8	96.04
Total 172		Total = 1833.60

$$S^2=\frac{\sum(x-x)^2}{n-1}=\frac{1833.60}{9}=203.73$$

$$H_0 : \mu = 100.; \ H_1 : \ \mu \neq 100$$

Test Statistic $t = \frac{\overline{x}-\mu}{\frac{S}{\sqrt{n}}} = \frac{97.2-100}{\sqrt{\frac{203.73}{1.0}}}$

$$t = -\ 0.6203$$

$$|t| = 0.62.$$

From table $t_{5\%}^{(9)} = 2.262$

Here $|t| = 0.62 < t_{5\%}^{(9)} = 2.262$

Hence null hypothesis is accepted.

The population mean I.Q is 100.

3

Variance Classification

3.1 Analysis of Variance: One Way Classification and Two Way Classification

Variance

Suppose that X is a random variable for the experiment, taking values in $S \subseteq R$, the expected value (or mean) of X gives the center of the distribution of X.

1. Variance and Standard Deviation:

The variance and standard deviation of X are defined by,

$$\text{var}(X) = E\left(\left[X - E(X)\right]^2\right)$$

$$\text{sd}\ (X) = \sqrt{\text{var}(X)}$$

The variance and standard deviation of X are both measures of the spread of the distribution about the mean. Variance is the square of that of X. Standard deviation, on the other hand, is the same as that of X. The standard deviation is often denoted by σ, so that the variance is σ^2.

The second moment of X about $a \in R$ is $E\ [(X-a)^2]$. Thus, the variance is the second moment of X about the mean $\mu = E(X)$ or equivalently, the second central moment of X. In general, the second moment of X about $a \in R$ can also be thought of as the mean square error if the constant a is used as an estimator of X.

In addition, if we think of the distribution of X as a mass distribution in R, then the second moment of X about $a \in R$ is the moment of inertia of the mass distribution about a. This is a measure of the resistance of the mass distribution to any change in its rotational motion about a. In particular, the variance of X is the moment of inertia of the mass distribution about the center of mass μ.

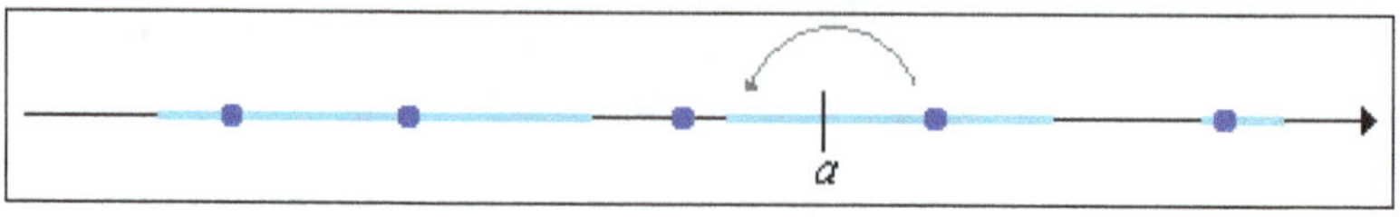

Moment of inertia about a.

2. Mean Square Error:

The mean square error (or equivalently the moment of inertia) about a is minimized when a = μ.

Let $mse(a) = E\left[(X - a)^2\right]$ for $a \in R$. Then mse is minimized when a = μ and the mini mum value is σ^2.

Proof:

Using the linearity of expected value:

$$mse(a) = E(X^2 - 2aX + a^2) = E(X^2) - 2aE(X) + a^2$$

Thus, the graph of mse is a parabola opening upward, with vertex at a = E(X).

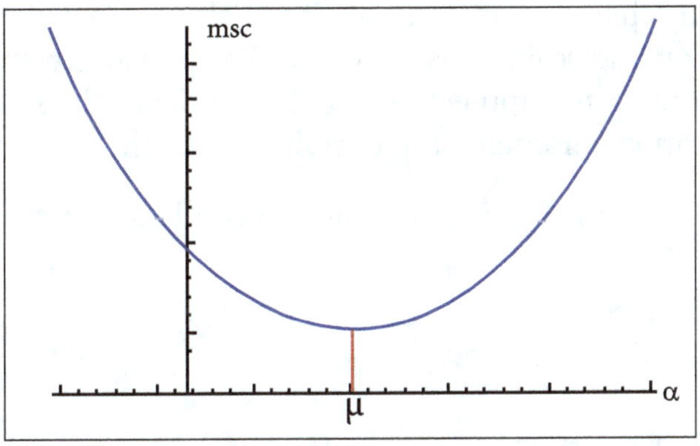

Graph of mse with vertex at a = E(X).

3. Variance Formula:

$$var(X) = E(X^2) - [E(X)]^2$$

Variance is always non-negative, since it has the expected value of non-negative random variable. Moreover, any random variable that really is random will have strictly positive variance.

4. The Non-Negative Property:

- var (X) ≥ 0
- var (X) = 0 if and only if P(X=c)=1 for some constant c (and then of course, E(X) = c).

Proof:

These results follow from the basic positive property of expected value.

5. Let $\mu = E(X)$. First $(X-\mu)^2 \geq 0$ with probability 1 so $E\left[(X-\mu)^2\right] \geq 0$. In addition, $E\left[(X-\mu)^2\right] = 0$ if and only if $P(X=\mu)=1$.

$$\text{var}(a + bX) = b^2 \ \text{var}(X)$$

$$\text{sd}(a + bX) = |b| \ \text{sd}(X)$$

Proof:

Let $\mu = E(X)$.

By linearity, $\text{var}(a+bX) = E\left(\left[(a+bX)-(a+b\mu)\right]^2\right) = E\left(b^2(X-\mu)^2\right) = b^2\text{var}(X)$.

For example, the change from inches to centimeters in a measurement of length is a scale transformation and the change from Fahrenheit to Celsius in a measurement of temperature is both the location and scale transformation. The previous result shows that when a location-scale transformation is applied to a random variable, the standard deviation does not depend on the location parameter, but is multiplied by the scale factor.

Let $\mu = E(X)$ and $\sigma^2 = \text{var}(X)$. A particularly important location-scale transformation of X is given by:

$$Z = \frac{X-\mu}{\sigma}$$

Z is called the standard score of X.

6. Let Z Denote the Standard Score of X. Then:

- $E(Z) = 0$
- $\text{var}(Z)=1$

Proof:

From the linearity of expected value, $E(Z) = 1/\sigma\left[E(X)-\mu\right] = 0$

From the scaling property, $\text{var}(Z) = 1/\sigma^2\text{var}(X) = 1$.

Since X and its mean and standard deviation have the same physical units, the standard score Z is dimensionless. It measures the directed distance from E(X) to X in terms of standard deviations.

7. Let Z denote the standard score of X and suppose that $Y = a + bX$ where a, b$\in$R and b$\neq$0.

- If $b > 0$, the standard score of Y is Z.

- If b < 0, the standard score of Y is −Z.

Proof:

$$E(Y)=a+bE(X) \text{ and } sd(Y)=|b|sd(X). \text{ Hence:}$$

$$\frac{Y-E(Y)}{sd\,(Y)}=\frac{b}{|b|}\frac{X-E(X)}{sd(X)}$$

When b > 0, the variable $Y=a+bX$ is a location-scale transformation and often corresponds to a change of physical units. Since the standard score is dimensionless, it's reasonable that the standard scores of X and Y are the same. On the other hand, when X ≥ 0, the ratio of standard deviation to mean is called as the coefficient of variation. This quantity is also dimensionless and is sometimes used to compare the variability for random variables with different means.

$$cv(X)=\frac{sd(X)}{E(X)}$$

Skewness

Suppose that X is a real valued random variable for the experiment. The mean of X is a measure of the center of the distribution of X. Furthermore, the variance of X is the second moment of X about the mean and measures the spread of the distribution of X about the mean. The third and fourth moments of X about the mean also measures the interesting features of distribution. The third moment measures skewness, the lack of symmetry, while the fourth moment measures kurtosis, the degree to which the distribution is peaked.

The actual numerical measures of these characteristics are standardized to eliminate by dividing by an appropriate power of the standard deviation. We assume that all the expected values given below exist and we will let μ = E(X) and σ2 = var(X). We assume that σ > 0, so that the random variable is really random.

1. The skewness of X is the third moment of the standard score of X.

$$skew\,(X)=E\left[\left(\frac{X-\mu}{\sigma}\right)^3\right]$$

The distribution of X is said to be positively skewed, negatively skewed or unskewed depending on whether skew(X) is positive, negative or 0.

In the unimodal case, if the distribution is positively skewed then the probability density function has a long tail to the right and if the distribution is negatively skewed, then the probability density function has a long tail to the left. A symmetric distribution is unskewed.

2. If the distribution of X is symmetric about a. That is, the distribution of a–X is the similar to the distribution of X–a. Then:

- E(X) = a
- Skew (X) = 0.

3. skew(X) can be described in terms of the first three moments of X.

$$\text{skew}(X) = \frac{E(X^3) - 3\mu E(X^2) + 2\mu^3}{\sigma^3} = \frac{E(X^3) - 3\mu\sigma^2 - \mu^3}{\sigma^3}$$

Proof:

Note that $(X-\mu)_3 = X_3 - 3X_2\mu + 3X\mu_2 - \mu_3$. From the linearity of expected value, we have:

$$E\left((X-\mu)^3\right) = E(X^3) - 3\mu E(X^2) + 3\mu^2 E(X) - \mu^3 = E(X^3) - 3\mu E(X^2) + 2\mu^3$$

The second expression follows from substituting the $E(X^2) = \sigma^2 + \mu^2$.

Since the skewness is referred in terms of an odd power of the standard score, it's invariant under a linear transformation with positive slope. On the other hand, if the slope is negative, skewness changes sign.

4. Suppose that $a \in R$ and $b \in R|\{0\}$. Then:

- $\text{skew}(a + bX) = \text{skew}(X)$ if $b > 0$
- $\text{skew}(a + bX) = -\text{skew}(X)$ if $b < 0$

Analysis of variance one way classification and two way classification:

- In one–way classification the data are categorized according to only one criterion.
- In a two–way classification the data are classified according to two different criteria.

Completely Randomized Design

Consider an experiment of agriculture in which "h" treatments and "n" plots are available. To control the extraneous variables treatment "I" should be replicated on "n I" plots, treatment 2 must be replicated on "n2" plots and so on. To reduce the error we have to randomize this process that is which nl plots be used treatment I and so on. For this we number the plots (from I to n) and write the numbers on the cards and shuffle well. Now, we select nl cards on which treatments we will be used and so on. This process and design is called as completely randomized Design.

Advantages

- Complete flexibility is allowed - any number of treatments and replicates might be used.
- Relatively easy statistical analysis, even with variable replicates and variable experimental errors for the different treatments.
- Analysis remains simple when data are missing.
- Provides the maximum number of degrees of freedom for error for a given number of experimental units and treatments.

Disadvantages

- Relatively low accuracy due to the lack of restrictions which allows environmental variation to enter experimental error.
- Not suited for large numbers of treatments because a relatively large amount of experimental material is needed which increases the variation.

S. No.	Randomized Block Design (RBD)	Completely Randomized Design (CRD)
1.	It has a simple layout but it is more efficient than CRD because of reduction of experimental error.	It has a simple layout.
2.	It is flexible and so any number of treatments and any number of replication may be used.	There is complete flexibility as the number of replication is not fixed.
3.	The analysis of the design is simple as it results in a two-way classification analysis of variance.	The analysis of the design is simple as it results in a one way classification analysis of variance.

Problems

1. A completely randomized design experiment with 10 plots and 3 treatments gave the following results.

Plot No:	1	2	3	4	5
Treatment:	ABC			A	C
Yield:	5	4	3	7	5
Plot No:	6	7	8	9	10
Treatment:	CAB			A	B
Yield:	1	3	4	1	7

Analyze the results for treatment effects.

Solution:

Given:

All the plots are not applied with the same number of treatments and we have to analyze the treatment effects only, we apply One way classification.

Re arranging the data,

Treatment A: 5 7

Treatment B: 4 4

Treatment C: 3 5

Step 1: Null Hypothesis

H_o: There is no difference in the yield between treatments.

Step 2: Test Statistic

5	4	3
7	4	5
3	7	1
1		
16	15	9

Total = T = 16+15+9 = 40

The square of the values are:

25	16	9
49	16	25
9	49	1
1		
84	81	35

$$LX^2 = 84 + 81 + 35 = 200$$

Correction factor = 160

$$SST = 40$$

$$SSC = (16)2 + (15)2 + (9)2 \ -160 = 6$$

$$SSE = SST - SSC = 40 - 6 = 34$$

The ANOVA table is:

Source of Variation	Sum of Square	Degrees of Freedom	Mean Sum of Squares	F -Ratio
Between Brands	SSC = 6	No. of Sample. 1= 3-1 =2	MSC $\frac{6}{3/2}$	$\frac{4.86}{3}=1.62^*$
Within Brands	SSE = 34	No. of Data – No. of Samples = 10-3 =7	MSE=$\frac{34}{7}$=4.86	

Step 3: Level of significance: Not given. So, let us take S % = O.OS

Step 4: Degrees of Freedom: (7,2)

Step 5: Table value: Fo. os(7,2)= 19.3S

Step 6: Conclusion: Comparing the F - value calculated at step 2 with table value of step S Fca1 < Flab, we accept the Null hypothesis. That is there is no difference in the mean yield between the treatments.

Assumptions involved in ANOVA:

- The observations are independent.
- Parent population from which observations are taken is normal.
- Various treatment and environmental effects are additive in nature.

2. Let us carry out ANOVA for the following data:

Workers	A	B	C	D
1	44	38	47	36
2	46	40	52	43
3	34	36	44	32
4	43	38	46	33
5	38	42	49	39

Solution:

Given:

- H_0: The mean productivity is the same for 4 difference machines.
- The 5 men do not differ with respect to mean productivity.

The above value subtracting 40 from each value.

Workers	Machine Type				Total
	A	B	C	D	
1	4	−2	7	−4	5
2	6	0	12	3	21
3	−6	−4	4	−8	− 14
4	3	−2	6	−7	0
5	−2	2	9	− 1	8
	5	−6	38	− 17	20

Step 1: $T = 20$

Step 2: $C.F = \frac{T^2}{N} = \frac{400}{20}$

$= 20$

Step 3: $SSR = \frac{\left(\sum y_1\right)^2}{n_1} + \frac{\left(\sum y_2\right)^2}{n_1} + \frac{\left(\sum y_3\right)^2}{n_1} + \ldots - C.F$

($\therefore n_1$ is the number of elements in each row)

$= \frac{5^2}{4} + \frac{21^2}{4} + \frac{(-14)^2}{4} + \frac{0^2}{4} + \frac{8^2}{4} - 20$

$= 161.5$

Degree of freedom, $v_2 = r - 1$

$= 5 - 1$

$= 4$

Step 4: $SSC = \frac{\left(\sum x_1\right)^2}{n} + \frac{\left(\sum x_2\right)^2}{n} + \frac{\left(\sum x_3\right)^2}{n} + \frac{\left(\sum x_4\right)^2}{n} - C.F$

$= \frac{5^2}{5} + \frac{(-6)^2}{5} + \frac{(38)^2}{5} + \frac{(-17)^2}{5} - 20$

$= 338.8$

Degree of freedom, $y_1 = k - 1$

$= 4 - 1$

$= 3$

Step: 5 $TSS = \text{Sum of squares ofeach values} - C.F$

$$= 4^2 + 2^2 + 7^2 + 4^2 + 6^2 + 0^2 + 12^2$$

$$+13^2 + 6^2 + 4^2 + 4^2 + 8^2 + 3^2 + 2^2$$

$$+ 6^2 + 7^2 + 2^2 + 2^2 + 9^2 + I^2 - 20$$

$$= 594 - 20$$

$$= 574$$

B 19:

$$SSE = TSS - (SSC + SSR)$$

$$= 574 - 338.8 - 161.5$$

$$= 73.7$$

$$v_3 = (k-1)(r-1) = 3 \times 4 = 12$$

ANOVA Table

Source of variations	Sum of Squares	Degree of freedom	Mean squares	F. Ratio
Between Columns	SSC = 338.8	$k - 1$ $= 4 - 1$ $= 3$	$MSC = \frac{SSC}{d.f}$ =112.933	$F_C = \frac{MSC}{MSE}$ =18.38
Between rows	SSR = 161.5	$r - 1$ $= 5 - 1$ $= 4$	$MSR = \frac{SSR}{d.f}$ =40.375	$F_R = \frac{MSR}{MSE}$ =6.574
Residual	SSE = 73.7	$(k-1)(r-1)$ = 12	$MSR = \frac{SSR}{d.f}$ =6.142	

(i) For v, = 3, v_3 = 12, $F_{v1', v}$ = 349 = F_t at 5%

Calculation value F_c = 18.38 > F_t = 3.49

∴ Mean productivity is not the same for the 4 different types of machines.

(ii) For v_2 = 4, v_3 = 12, $F_{4, 12}$ = 3.26 = F_t at 5%

The computed value $F_c = 6.58$ is greater than the table value $F_t = 3.26$.

∴ The workers differ with respect to mean productivity.

3.2 Latic Square Design

Shape of the field should be approximately square or rectangular called Latin square. Suitable mainly for 5 (or) 10 treatments.

The Latin Square Designs are not used probably as much as they should be - they are very efficient designs. The Latin square designs allow for 2 blocking factors. In other words, these designs are used to simultaneously control the (or eliminate) 2 sources of nuisance variability. For the instance, if we had a plot of land the fertility of this land might change in both directions, North - South and East - West due to soil or moisture gradients. So, both rows and columns can be used as blocking factors. But, we can use Latin squares in lots of other settings. As we shall see, Latin squares can be employed as much as the RCBD in industrial experimentation as well as other experiments.

Whenever, we have more than one blocking factor a Latin square design will enable we to remove the variation for these 2 sources from the error variation. So, consider we had a plot of land, we might have blocked it in columns and rows, that is each row is a level of the row factor and each column is a level of the column factor. We can eliminate the variation from our measured response in both directions if we take both the rows and columns as factors in our design.

The Latin Square Design obtain its name from the fact that we can write it as a square with Latin letters to correspond to the treatments. The treatment factor levels are the Latin letters in the Latin square design. The number of rows and columns has to correspond to the number of treatment levels. So, if we have 4 treatments then we would require having four rows and four columns in order to create a Latin square. This provides us a design where we have each of the treatments and in each row and in each column.

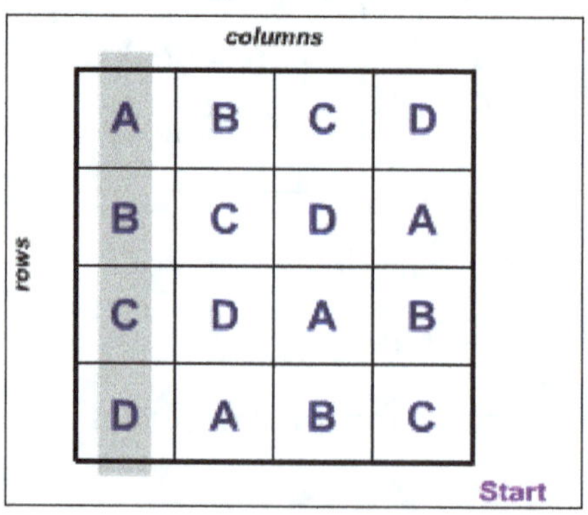

This is just one of several 4×4 squares that we could create. In fact, we can make any size square we want, for any number of treatments - it just requires to have the following property associated with it - that each treatment occurs only once in each row and once in each column.

Let us consider another example in an industrial setting: the rows are the batch of raw material, the columns are the operator of the equipment and the treatments (A, B, C and D) are an industrial process or protocol for producing a particular product.

We let:

$$y_{ijk} = \mu + \rho_i + \beta_j + \tau_k + e_{ijk}$$

$$i = 1, \ldots, t$$

$$j = 1, \ldots, t$$

$[k = 1, \ldots, t]$ Where - $k = d(i, j)$ and the total number of observations,

$N = t^2$ (the number of rows times the number of columns) and t is the number of treatments.

Note that the Latin Square is an incomplete design, which means that it does not include the observations for all possible combinations of i, j and k. This is why we use notation $k = d(i, j)$. Once we know the row and column of the design, then the treatment is declared. In other words, if we know i and j, then k is specified by the Latin Square design.

This property has an impact on how we determine the means and sums of squares and for this reason we cannot use the balanced ANOVA command in Minitab even though it looks perfectly balanced.

An assumption that we make when using a Latin square design is that the 3 factors (treatments and 2 nuisance factors) do not interact. If this kind of assumption is violated, the Latin Square design error term will be inflated.

The randomization method for assigning treatments that we would like to use when we actually give a Latin Square is somewhat restricted to preserve the structure of the Latin Square. The ideal randomization would be to select a square from the set of all possible Latin squares of the specified size. However, a more practical randomization scheme would be to select a standardized Latin square at random (these are tabulated) and then:

- Randomly permute the columns.
- Randomly permute the rows and then.
- Assign the treatments to the Latin letters in a random fashion.

Let us consider a factory setting where we are producing a product with 4 operators and 4 machines. We call the columns the operators and the rows the machines. Then we can randomly assign the specific operators to a row and the specific machines to a column. The treatment is one of 4 protocols for producing the product and our interest is in the average time needed to produce each product. If both the machine and the operator have an effect on the time to produce, then by using a Latin Square Design this variation due to machine or operators will be effectively removed from the analysis.

The following table provides the degrees of freedom for the terms in the model.

AOV	df	df for the example
Rows	$t-1$	3
Cols	$t-1$	3
Treatments	$t-1$	3
Error	$(t-1)(t-2)$	6
Total	(t^2-1)	15

A Latin Square design is actually simple to analyze. Because of the restricted layout, one observation per treatment in each row and column, the model is orthogonal.

If the row, ρ_i and column, β_j, effects are random with expectations zero, the expected value of Y_{ijk} is $\mu + \tau_k$. In other words, the treatment effects and treatment means are orthogonal to the row and column effects. We can also write the sums of squares, as seen in Table above.

We can test for row and column effects, but our focus of interest in a Latin square design is on the treatments. Just as in RCBD, the row and column factors are included to decrease the error variation but are not usually of interest. And, depending on how we've conducted the experiment they often have not been randomized in a way that allows us to create any reliable inference from those tests.

Note: if we have missed the data then we need to use the general linear model and test the effect of treatment after fitting the model that would account for the row and column effects.

In general, the General Linear Model tests the hypothesis that:

$$H_o: \tau_i = 0 \text{ vs } H_a: \tau_i \neq 0$$

To test this hypothesis we will look at the F-ratio which is written as:

$$F = \frac{MS(T_k / \mu, \rho_i, \beta_j)}{MSE\ (\mu, \rho_i, \beta_j, T_k)} \sim F((t-1), (t-1)(t-2))$$

To get this in Minitab we would use GLM and fit the 3 terms: rows, columns and treatments. The F statistic is based on the adjusted MS for treatment.

Statistical Analysis of the Latin Square Design

The statistical (effects) model is:

$$Y_{ijk} = \mu + \rho_i + \beta_j + T_k + \varepsilon_{ijk} \begin{cases} i = 1,2 \ldots. p \\ j = 1,2 \ldots. p \\ k = 1,2 \ldots. p \end{cases}$$

But k = d (i, j) represents the dependence of k in the cell i, j on the design layout and p = t the number of treatment levels.

The statistical analysis (ANOVA) is much like the analysis for the RCBD.

Advantages

The benefits of Latin square designs are:

- They handle the case when we have several nuisance factors and we either can't combine them into a single factor or we wish to keep them separate.
- They allow experiments with a relatively small number of runs.
- Latin square controls more of the variation than the completely randomized design or the randomized complete block design. The two–way elimination of variation often results in small error mean square.
- The analysis is simple; it is only slightly more complicated than that for the randomized complete block design.
- The analysis remains relatively simple even with missing data. Analytical procedures are available after omitting one or more treatments, rows or columns.

Disadvantages

Drawbacks:

- The number of levels of each blocking variable should equal the number of levels of the treatment factor.
- The Latin square model assumes that there are no interactions between the blocking variables or between the treatment variable and the blocking variable.
- The number of treatments must equal the number of replicates.
- The experimental error is likely to increase with the size of the square.

- Small squares have very few degrees of freedom for experimental error.
- We can't evaluate interactions between:
  - Rows and columns.
  - Rows and treatments.
  - Columns and treatments.

Effect of the Size of the Square on Error Degrees of Freedom:

SOV	Df	2 × 2	3 × 3	4 × 4	5 × 5	8 × 8
Rows	r-1	1	2	3	4	7
Columns	r-1	1	2	3	4	7
Treatments	r-1	1	2	3	4	7
Error	(r-1) (r-2)	0	2	6	12	42
Total	r^2-1	3	8	15	24	63

Where r = number of rows, columns and treatments.

One way to increase the Error df for small squares is to use more than one square in the experiment (i.e. repeated squares).

Problems

1. Let us perform Latin square experiment for the following.

Roma	I	II	III	→	Three equally spaced concentrations of poison as extracted from the scorpion fish.
Arabic	1	2	3	→	Three equally spaced body weights for the animals tested.
Latin	A	B	C	→	Three equally spaced times of storage of the poison before it is administered to the animals.

	I	II	III
1.	0.194	0.73	1.187
	A	B	C
2.	0.758	0.3111	0.589
	C	A	B
3.	0.369	0.558	0.311
	B	C	A

Solution:

Given:

H_0: There is no significant difference between rows, between columns and treatments.

A (0.194)	B (0.73)	C (1.18)
C (0.758)	A (0.311)	B (0.589)
B (0.369)	C (0.558)	A (0.311)

To find: TSS, SSR & SSC

	C_1	C_2	C_3	Row total Ri	$R_i^{2/3}$
R_1	0.194	0.73	1.187	2.111	1.485
R_2	0.758	0.311	0.589	1.658	0.916
R_3	0.369	0.558	0.311	‘ 1.238	0.511
C_j	1.321	1.599	2.087	5.007	2.912
				T	
$C_j^{2/3}$	0.582	0.852	1.452	2.886	–

To find: SST

	1	2	3	Row Total T_i	$T_i^{2/3}$
A	0.194	0.311	0.311	0.816	0.2219
B	0.369	0.73	0.589	1.688	0.9498
C	0.758	0.558	1.187	2.503	2.0883
					3.2600

- $T = 5.007$
- $C.F = \frac{T^2}{N} = \frac{(5.007)^2}{9} \Rightarrow 2.79$
- Square of individual Observations

$$= (0.194)^2 + (0.73)^2 + (1.187)^2 + (0.758)^2$$

$$+ (0.311)^2 + (0.589)^2 + (0.369)^2$$

$$+ (0.558)^2 + (0.3\,ll)^2$$

$$= 3.542$$

- $\text{TSS} = 3.542 - 2.79$

  $= 0.752$

- SSR = Sum of squares of rows

  $= \sum\left(\frac{R_i^2}{3}\right) - \text{C.F} = 2.912 - 2.79$

  $= 0.122$

- SSC – Sum of squares of columns

  $= \sum\left(\frac{C_j^2}{3}\right) - \text{C.F} = 2.886 - 2.79$

  $= 0.096$

- SST = Sum of squares of Treatments

  $= \sum\left(\frac{T_i^2}{3}\right) - \text{C.F} = 3.26 - 2.79$

  $= 0.47$

- $\text{SSE} = \text{TSS} - (\text{SSR} + \text{SSC} + \text{SST})$

  $= 0.752 - (0.122 + 0.096 + 0.47)$

  $= 0.064$

ANOVA Table

Source of Variations	Sum of Squares	difference	Mean squares	F-ratio
Rows	SSR = 0.122	3−1=2	$\text{MSR} = \frac{0.122}{2} = 0.061$	$F_R = \frac{\text{MSR}}{\text{MSE}} = 1.906$
Columns	SST = 0.096	3−1=2	$\text{MSC} = \frac{0.096}{2} = 0.048$	$F_C = \frac{\text{MSE}}{\text{MSC}} = 1.58$
Treatments	SST = 0.47	3−1=2	$\text{MST} = \frac{0.47}{2} = 0.235$	$F_t = \frac{\text{MST}}{\text{MSE}} = 7.34$
Residual	SSE = 0.064	(3−1) (3−2) = 2	$\text{MSE} = \frac{0.064}{2} = 0.032$	

Table value of F for (2,2) degree of freedom at 5% level of significance is 19.00.

i.e., $F_{tab} = 19.00$

Since $F_R < F_{tab}$

$\therefore$ Ho is accepted.

i.e., there is no significant difference between rows.

Table value of F for (6,3) degree of freedom at 5% level of significance is 8.94.

$F_{tab} = 8.94$

Since $F_c < F_{tab}$: Ho is accepted.

i.e., There is no significant difference between columns. Since $F_T > F_t$ H_o is rejected.

i.e., There is no significant difference between treatments.

2. A variable trial was conducted on wheat with 4 varieties in a Latin Square design. The plan of the experiment and per plot yield are given below:

C25	B23	A20	D20
A19	D19	C21	B18
B19	A14	D17	C20
D17	C20	B21	A15

Let us analyze the data.

Solution:

Given:

Subtract 18 from the given values and work out with the new values of x_{ij}:

j / i	1	2	3	4	T_i $\frac{T_i^2}{n}$	T_i $\frac{T_i^2}{n}$	$\sum_i x_{ij}^2$
1	C7	B5	A2	D2	16	64	82
2	A1	D1	C3	Bo	5	6.25	11
3	B1	A–4	D–l	C2	–2	1	22
4	D–l	C2	B3	A–3	1	0.25	23
T_j	8	4	7	1	T = 20	T=20 $\sum\left(\frac{T_i^2}{n}\right)$=71.5	138

$\frac{T_j^2}{n}$	16	4	12.25	0.25	$\sum(T_j^2/n)$ =32.5		
$\sum_i x_{ij}^2$	52	46	23	17	138		

Re–arranging the data according to the letters:

Letter	X_k				T_k	$\frac{T_k^2}{n}$
A	2	1	−4	−3	−4	4
B	5	0	1	3	9	20.25
C	7	3	2	2	19	49
D	2	1	−1	−1	1	0.25
Total					20	73.5

$$Q=\sum\sum x_{ij}^2-\frac{T^2}{N}=138-\frac{(20)^2}{16}=113$$

$$Q_1=\frac{1}{n}\sum T_i^2-\frac{T^2}{N}=71.5-25=46.5$$

$$Q_2=\frac{1}{n}\sum T_j^2-\frac{T^2}{N}=32.5-25=7.5$$

$$Q_3=\frac{1}{n}\sum T_k^2-\frac{T^2}{N}=73.5-25=48.5$$

$$Q_4=Q-Q_1-Q_2-Q_3=113-(46.5+7.5+48.5)$$

$$=10.5$$

ANOVA Table

Source of Variation (S.V)	Sum of Squares (S.S)	Degree of Freedom d.f	Mean Square (M.S)	Variance Ratio
Between Rows	Q1 = 46.5	n – 1=3	15.5	F_R = 8.857
Between Columns	Q2 =7.5	n – 1 = 3	2.5	F_c = 1.429
Between Letters	Q3 = 48.5	n – 1 = 3	16.167	F_l = 9.24
Residual	Q4 = 10.5	(n – 1) (n – 2) = 6	1.75	–
Total	Q = 113	(n2 – 1) = 15	–	–

From F–table, F5 % ($V_1 = 3, V_2 = 6$) = 4.76.

Between Letters

Since F_L(= 9.24) > F5 % (3, 6) (= 4.76) with respect to the letters; Hypothesis is rejected.

The difference between the methods of cultivation is significant.

3. Let us analyze the variance in the Latin square of yields (in kgs) of paddy where P, Q, R, S denote the different methods of cultivation:

S122	P121	R123	Q122
Q124	R123	P122	S125
P120	Q119	S120	R121
R122	S123	Q121	P122

Solution:

Given:

S122	P121	R123	Q122
Q124	R123	P122	S125
P120	Q119	S120	R121
R122	S123	Q121	P122

Subtract 120 from the given values and work out with the new values of x_{ij}:

j / i	1	2	3	4	Ti	$\frac{T_i^2}{n}$	$\sum_j x_{ij}^2$
1	S_2	P_1	R_3	Q_2	8	16	18
2	Q_4	R_3	P_2	S_5	14	49	54
3	P_0	Q_{-1}	S_0	R_1	0	0	2
4	R_2	S_3	Q_1	P_2	8	16	18
T_j	8	6	6	10	T = 30	$\frac{\sum T_i^2}{n} = 81$	$92 = \sum\sum x_{ij}^2$
$\frac{T_{ij}^2}{n}$	16	9	9	25	$\frac{\sum T_j^2}{n} = 59$		
$\sum_i x_{ij}^2$	24	20	14	34	$92 = \sum\sum x_{ij}^2$		

Re-arranging the data according to the letters. We have:

Letter	x_k				T_k	$\frac{T_k^2}{n}$
P	1	2	0	2	5	6.25
Q	2	4	-1	1	6	9
R	3	3	1	2	9	20.25
S	2	5	0	3	10	25
	Total				T = 30	$\sum\frac{T_k^2}{n}=60.5$

$$Q=\sum\sum x_{ij}^2-\frac{T^2}{N}=92-\frac{30^2}{16}=35.75$$

$$Q_1=\sum\frac{T_i^2}{n}-\frac{T^2}{N}=81-56.25=24.75$$

$$Q_2=\sum\frac{T_1^2}{n}-\frac{T^2}{N}=59-56.25=2.75$$

$$Q_1=\sum\left(\frac{T_k^2}{n}\right)-\frac{T^2}{N}=60.5-56.25=4.25$$

ANOVA Table

S.V	S.S	Degree of Freedom d.f	M.S	F_0
Between Rows	Q_1=24.75	n-1=3	$\frac{Q_1}{n-1}=8.25$	-
Between Columns	Q_2=2.75	n-1=3	$\frac{Q_2}{n-1}=0.92$	-
Between Letters	Q_3=4.25	n-1=3	$\frac{Q_3}{n-1}=1.42$	$\frac{1.42}{0.67}=2.12$
Residual	Q_4=4	(n-1) (n-2)=6	$\frac{Q_4}{(n-1)(n-2)}=0$	67
Total	Q = 35.75	n^2 - 1 = 15	-	-

Between Letters

From the F-Table, $F_{5\%}$ (V1 = 3, V2 = 6) = 4.76.

Here calculated F = 2.12 less than Tabulated F = 4.76 with respect to between the letters.

∴ The difference between the method of cultivation is not significant.

3.3 RBD

Rbd in the Sense Randomized Block Design

In a randomized block design, there is only one primary factor under consideration in the experiment. Similar test subjects are collected into blocks. Each block is tested against all treatment levels of the primary factor at random order. This is intended to delete possible influence by other extraneous factors.

$$\frac{S_t^2}{(k-1)}=\frac{SST}{(k-1)}=S_t^2 \text{ (MSS due to treatments)}$$

$$\frac{S_E^2}{(N-k)}=\frac{SSE}{(N-k)}=S_E^2 \text{ (MSS due to error)}.$$

Within randomized block designs, we have 2 factors:

- Blocks
- Treatments

A randomized complete block design with a treatments and b blocks is constructed in 2 steps:

- The experimental units (the units to which our treatments are going to be applied) are partitioned into b blocks, each composed of a units.
- Treatments are the randomly assigned to the experimental units in such a way that each treatment appears once in each block.

Randomized block designs are often applied in agricultural settings. The example below will make this clearer.

In general, the blocks must be partitioned so that:

- Units within blocks are as uniform as possible.
- Differences between blocks are as large as possible.
- These conditions will usually give us the most powerful results.

Experimental Design Example: Rice Data

We have 4 different varieties of rice; varieties A, B, C and D. And, we have 5 different blocks in our study. So, imagine each of these blocks as a rice field or patty on a farm somewhere. These blocks are just different patches of land and each block is partitioned into 4 plots. Then we randomly assign which variety goes into which plot in each block. We will note that variety A appears once in each block, as does each of the other types. This is how the randomized block design experiment is set up.

Block 1		Block 2		Block 3		Block 4		Block 5	
D	C	A	D	B	D	D	B	A	C
B	A	B	C	C	A	A	C	D	B

A randomized block design with the following layout was used to compare 4 varieties of rice in 5 blocks.

This kind of experimental design is also used in medical trials where we would be looking for people with similar characteristics to be in every block. This might be people who weigh about the same, are of the same sex, same age or whatever factor is deemed important for that particular experiment. So usually, what we want are people within each of the blocks to be similar to one another.

In each of the partitions within each of the 5 blocks one of the 4 varieties of rice would be planted. In this experiment the height of the plant and the number of tillers per plant were measured 6 weeks after transplanting. Both of these measurements are indicators of how vigorous the growth is. The taller the plant and the greater number of tillers, the healthier the plant is, which must lead to a higher rice yield.

In general, randomized block design data should look like this:

$$
\begin{array}{llcccc}
 & & & \text{Block} & & \\
 & & 1 & 2 & \cdots & b \\
\hline
 & 1 & Y_{11} - \begin{pmatrix} Y_{111} \\ Y_{112} \\ \vdots \\ Y_{11p} \end{pmatrix} & Y_{12} - \begin{pmatrix} Y_{121} \\ Y_{122} \\ \vdots \\ Y_{12p} \end{pmatrix} & \cdots & Y_{1b} - \begin{pmatrix} Y_{1b1} \\ Y_{1b2} \\ \vdots \\ Y_{1bp} \end{pmatrix} \\
\text{Treatment} & 2 & Y_{21} - \begin{pmatrix} Y_{211} \\ Y_{212} \\ \vdots \\ Y_{21p} \end{pmatrix} & Y_{22} - \begin{pmatrix} Y_{221} \\ Y_{222} \\ \vdots \\ Y_{22p} \end{pmatrix} & \cdots & Y_{2b} - \begin{pmatrix} Y_{2b1} \\ Y_{2b2} \\ \vdots \\ Y_{2bp} \end{pmatrix} \\
 & \vdots & \vdots & \vdots & & \vdots \\
 & a & Y_{a1} - \begin{pmatrix} Y_{a11} \\ Y_{a12} \\ \vdots \\ Y_{a1p} \end{pmatrix} & Y_{a2} - \begin{pmatrix} Y_{a21} \\ Y_{a22} \\ \vdots \\ Y_{a2p} \end{pmatrix} & \cdots & Y_{ab} - \begin{pmatrix} Y_{ab1} \\ Y_{ab2} \\ \vdots \\ Y_{abp} \end{pmatrix} \\
\hline
\end{array}
$$

We have rows for the treatments. In this case we would have 4 rows, one for each of the 4 varieties of rice. We also set up b columns for b blocks. In this case we have 5 columns, one for each of the 5 blocks. In each block, for each treatment we are going to observe a vector of variables.

Our notation is as follows:

Let Y_{ijk} = observation for variable k from block j in treatment i

We will then collect these into a vector Y_{ij} which looks like this:

$$Y_{ij} - \begin{pmatrix} Y_{ij1} \\ Y_{ij2} \\ \vdots \\ Y_{ijp} \end{pmatrix}$$

a = Number of Treatments

b = Number of Blocks

Problems

1. The cutting speeds of four types of tools are being compared in an experiment. Five cutting materials of varying degree of hardness are to be used as experimental blocks. The data giving the measurement of cutting time in seconds appear in the table below.

Treatment	Block					Treatment Means
	1	2	3	4	5	
1	12	2	8	1	7	$\bar{x}_{T1}=6$
2	20	14	17	12	17	$\bar{x}_{T2}=16$
3	13	7	13	8	14	$\bar{x}_{T3}=11$
4	11	5	10	3	6	$\bar{x}_{4}=7$
Block Means	$\bar{x}_{B1}=14$	$\bar{x}_{B2}=7$	$\bar{x}_{B3}=12$	$\bar{x}_{B4}=6$	$\bar{x}_{B5}=11$	$\bar{x}=10$

Solution:

Given:

The Randomized Block Design Model:

x_{ij} = the measurement for treatment i in block j (remember there is precisely one such measurement) then:

$$x_{ij}=\mu+\alpha_i+\beta_j+e_{ij}$$

Where, μ =overall mean

α_i = effect of treatment I (difference with μ).

β_j =effect of block j (difference with μ).

e_{ij} = error in measurement for treatment I and block j.

A positive value for αi indicates that the mean of the response variable is greater than the overall mean for treatment i.

Assume $e_{ij} \sim N(I,\sigma)$ for all measurements.

In the analysis of variance instead of only explaining the variance through error and treatment, we also include the block as a possible source for variance in the data. For that reason we now also include the following.

Sum of Squares for Block (SSB) with our analysis:

- Sum of Squares for treatment:

  $$SST = \sum_{i=1}^{k} b\left(\overline{x}_{T1} - \overline{x}\right)^2, \; df_T = k-1$$

- Sum of Squares for block:

  $$SSB = \sum_{j=1}^{b} k\left(\overline{x}_{B1} - \overline{x}\right)^2, \; df_B = b-1$$

- Total Sum of Squares:

  $$Total\,SS = \sum_{i,j} \left(x_{ij} - \overline{x}\right)^2, \; df_{Total} = n-1$$

- Sum of Squares for error:

  $$SSE = Total\ SS - SST - SSB, \; df_E = n = b - k + 1$$

- Summarized in an ANOVA-table:

ANOVA Table for a Randomized Block Design:

Source	df	SS	MS	F
Treatments	k-1	SST	MST = SST / (k-1)	MST/MSE
Blocks	b-1	SSB	MSB = SSB / (b-1)	MSB/MSE
Error	n-k-b+1	SSE	MSE = SSE / (n-k-b+1)	
Total	n-1	Total SS		

4

Statistical Quality Control

4.1 Concept of Quality of a Manufactured Product

1. The Quality is conformance to requirements or specifications.

This is the definition used by P. Crosby who suggests that in order to manage quality adequately, we must be able to quantify and measure it.

2. The Quality is fitness for use.

This is a more general definition attributed to J. Juran. It places emphasis on the consumer aspect of quality. This emphasizes that requirements and specifications translate fitness for use into measurable quantities.

3. The Quality is inversely proportional to variability.

It is the definition of D. Montgomery implies that if the variability in a product's important characteristics decreases the quality of the product increases. Thus, quality improvement is the reduction of variability in processes and products.

4. The Quality is the totality of features and characteristics of a product or service that bear on its ability to satisfy given needs.

It is the definition from the 1978 consensus definition in ANSI/ASQC Standard A3. If we replace the word given with stated or implied we have the ISO definition. – It implies that quality is an attribute of a product. A product is a quality product if it meets all the requirements established for it; that is, it is a defect-free product.

- The quality of a product can be evaluated in several ways referred to as the dimensions of quality. Therefore, it is major to differentiate among these dimensions:
    - Reliability: How often does the product fail?
    - Performance: Will the product do the intended job?
    - Serviceability: How easy is it to repair the product?
    - Durability: How long does the product last?
    - Features: What does the product do?

- Aesthetics: What does the product look like?
- Perceived quality: What is the reputation of the company or its product?
- Conformance to standards: Is the product made exactly as the designer intended?

5. Quality improvement is the reduction of variability in processes and products. One goal of this course is to learn how to implement statistical tools that can be used to improve quality.

6. For each product, the user or consumer defines certain properties of perceived quality referred to as quality characteristics.

7. Quality characteristics are usually related to specifications. For a manufactured product, specifications refer a range of acceptable measurements for the quality characteristics of the product components.

8. The expected value for a quality characteristic is called the nominal or target value. A target value is usually bounded by an acceptable interval of the values which will not impact the performance of the product if the quality characteristic value is in that range.

9. The largest allowable value for the quality characteristics is called the upper specification limit (USL) and the smallest allowable value for a quality characteristics is called the lower specification limit (LSL). Those specification limits are usually set by the product's end user.

10. A quality characteristic might only have one specification limit (USL or LSL, but not both).

11. A nonconforming product is a product that fails to meet one or more of its specifications. A specific type of failure is called non-conformity.

12. A nonconforming product is not necessarily unfit for use. For example, a cosmetic nonconformity can't affect the fitness of the product for use.

13. A nonconforming product is considered defective if it has one or more defects. A defect is a nonconformity that is serious enough to significantly affect the proper usage of the product.

Management Aspects of Quality

Quality Planning

- This planning is the development of strategic activities designed to improve the quality of a product. The planning can include both statistical and management activities.

Quality Assurance

- Another terminology related to the overall approach to quality control is quality assurance. It is designed to include checks and controls on the quality control method in use.
    - For a specific product or service, this might involve verification audits and the evolution of the quality factors that affect the specification, production, design, installation, procurement, test and inspection, sales, marketing and use of the product or service.
    - The objective of quality assurance is to ensure which quality control is doing what it must be doing. Such an effort should be organized within a factory, company or corporation and motivated by management.
- Quality Assurance is a system of activities whose purpose is to give an assurance that the overall quality control is in fact being done effectively.

Quality Control

- It is the regulatory process through which we measure actual quality performance.

  This is the definition provided by Juran (1974). It puts quality into an action mode that creates use of standards and measurements.

The Ansi/Asqc Standard A3 Definition (1978)

- The Statistical Control of Quality is application of statistical principles and methods in all stages of design, production, maintenance and service, directed toward the economic satisfaction of demand.

  The definition was given by Deming (1971) emphasizes the statistical applications and it also makes explicit the economic goals of quality control.

- Statistical process control (SPC) will be considered as statistical quality control applied to a process or to a product resulting from a process. The DuPont Co. definition of SPC is:

  SPC is the totality of all process activities directed at improving process consistency through detecting changes in measured characteristics, identifying the causes of changes and preventing recurrence of those causes.

Benefits of Quality Control in Industry

- Improving the quality of products.
- Reducing manufacturing and corporate costs.

- Increasing the productivity of manufacturing processes.
- Reducing consumer prices of products.
- Determining and improving the marketability of products.
- Improving and/or assuring on-time deliveries and availability.

Control Charts

A control chart provides a basis for deciding whether the variation in the output is due to common causes (in control) or assignable causes (out of control). Whenever an out-of-control situation is detected, adjustments and/or other corrective action can be taken to bring the process back into control.

Control charts will be classified by the type of data they contain. An $\bar{X}$ chart is used if the quality of the output is measured in terms of a variable such as length, weight, temperature and so on. In which case, the decision to continue or to adjust the production process can be based on the mean value found in a sample of the output. To introduce some of the concepts common to all control charts, let us consider some specific features of an $\bar{X}$ chart.

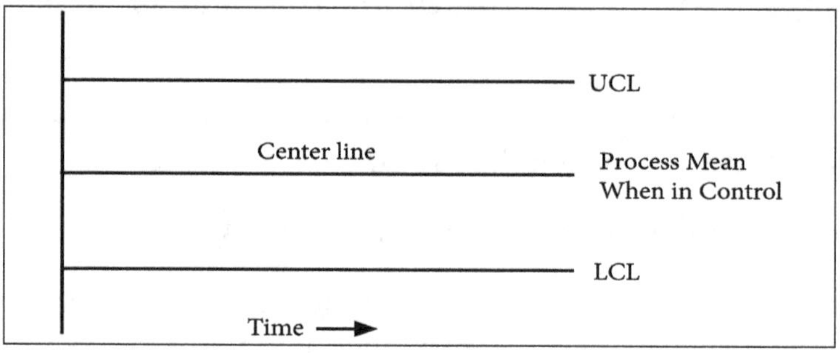

Figure shows the general structure of an $\bar{X}$ chart. The center line of the chart corresponds to the mean of the process when the process is in control. The vertical line identifies the scale of measurement for the variable of interest.

Outcome of the Statistical Process Control

		State of Production Process	
		H_0 True Process in Control	H_0 False Process Out of Control
	Continue Process	Correct decision	Type II error (allowing an out – of – control process to continue)
Decision	Adjust Process	Type I error (adjusting an in –control process)	Correct decision

Each time a sample is taken from the production process, a value of the sample mean x is computed and a data point showing the value of x is plotted on the control chart. The two lines labeled UCL and LCL are important in determining whether the process is in control or out of control. The lines are called the upper control limit and the lower control limit, respectively. They are chosen so that once the process is in control, there can be a high probability that the value of 'Twill be between the two control limits. Values outside the control limits give the strong statistical evidence that the process is out of control and corrective action must be taken. Over time, more and more data points can be added to the control chart. The order of the data points will be from left to right as the process is sampled. In essence, every time a point is plotted on the control chart, we are carrying out a hypothesis test to find out whether the process is in control. In addition to the x chart, other control charts will be used to monitor the range of the measurements in the sample (R chart), the proportion of defective items in the sample (p chart) and the number of defective items in the sample (np chart). In each case, the control chart has a LCL line, a center line and a UCL line same as the $\overline{x}$ chart in Figure. The major difference among the charts is what the vertical axis measures and for in-stance, in a p chart the vertical axis denotes the proportion of defective items in the sample instead of the sample mean.

4.1.1 Defects and Defectives

Customers expect the products and services to meet their specifications. When they do not, a defect or defective is present.

Defects

A defect may be any item or service that exhibits a departure from specifications. A defect does not necessarily mean that the product or service can't be used. A defect denotes only that the product result is not entirely as intended.

If service in a restaurant is being evaluated. If a waiter greets his table after five minutes, the customer can still order and enjoy the meal even though the promptness of the greeting did not meet expectations. Therefore, this could be considered a defect ("late greeting") in the service.

Defectives

A defective is an item or the service that is considered completely unacceptable for use. Every item or service experience is either considered defective or not there are only 2 choices.

Before final shipment, a quality inspector evaluates auto supply parts and rates each item as "pass" or "fail" to ensure that the company does not ship any parts that will be unusable.

Comparison of Defects and Defectives

A defective item has one or more defects. However, not all items with defects are defective. It depends on the severity of the defect. New cars might have many defects, some of which might not even be noticed by the customer. However, if the car has a defect that is measured and reported, the car (or part of the car) might be considered defective.

Let us consider the loan application process. In this case, the processing department is the customer. The form has 36 entries. We sample 50 forms to estimate the defect rate. One application has 7 incorrect entries—there are 7 defects present on this form. Another application has 4 incorrect entries—there are four defects present on this form.

Overall, 18 forms have at least one defect, so 18 forms are defective out of 50. Overall, there were 62 total defects per 1800 opportunities (36 opportunities per form * 50 forms).

Analyses for Defects and Defectives Data

The kind of statistical analysis that we use depends on whether we are evaluating defects or defectives:

To find out the defectives, we use analyses that are based on a binomial probability model, such as a 1 Proportion test, a 2 Proportions test, a P chart, an NP chart or a binomial capability analysis. These analyses evaluate the proportion of defectives in the process.

To evaluate defects, we use analyses that are based on a Poisson probability model, such as a 1-Sample Poisson rate test, a 2-Sample Poisson rate test, a C chart, a U chart or a Poisson capability analysis. These analyses determine the rate of defects in the process.

4.1.2 Causes of Variations

A phenomenon will be said to be controlled when, through the use of past experience, we can predict, at least within limits, how the phenomenon might be expected to differ in the future. Here it is understood that prediction within limits means that we can state, at least approximately, the probability that the observed phenomenon will fall within the provided limits.

The critical point in this definition is that control is not defined as the complete absence of variation. Control is simply a state where all variation is predictable variation. A controlled process is not necessarily a sign of good management, nor is an out-of-control process necessarily producing non-conforming product.

In every of prediction there is an element of chance. For our usages, we will call any unknown random cause of difference a chance cause or a common cause, the terms are synonymous and will be used as such. If the influence of any particular chance cause is very small and if the numbers of chance causes of variation are very large and relatively constant, we have a situation where the variation is predictable within limits. We can see from the definition above, that a system such as this qualifies as a controlled system. Where Dr. Shewhart used the term chance cause, Dr. W. Edwards Deming coined the term general cause to describe the same phenomenon.

Common Causes

- Poor design.
- Inappropriate procedures.
- Lack of clearly defined standard operating procedures.
- Poor maintenance of machines.
- Machines not suited to the job.
- Poor working conditions, Example: Lighting, noise, dirt, temperature, ventilation.
- Substandard raw materials.
- Measurement error.
- Vibration in industrial processes.
- Ambient temperature and humidity.
- Insufficient training.
- Normal wear and tear.
- Variability in settings.
- Computer response time.

Special Causes

- Operator absent.
- Poor adjustment of equipment.
- Operator falls asleep.
- Faulty controllers.
- Machine malfunction.
- Computer crashes.

- Poor batch of raw material.
- Power surges.

4.1.3 Random and Assignable

The Simple random sampling is the most widely-used probability sampling technique, probably because it is easier to implement and easy to analyze.

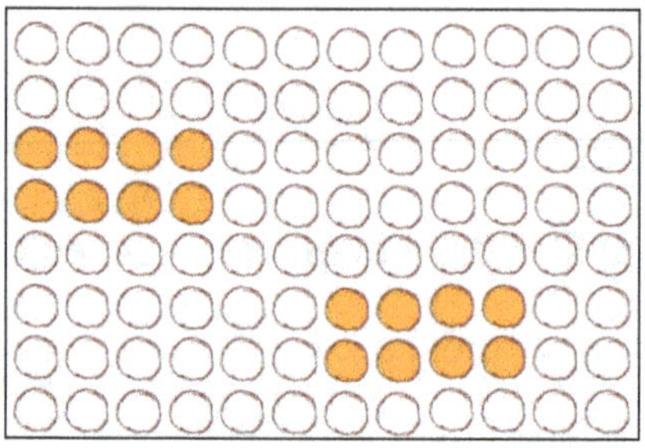

Key Definitions

To understand the simple random sampling, first we need understand a few key definitions:

- The total set of observations that can be made is called the population.
- A sample is a collection of observations drawn from a population.
- A parameter is a measurable characteristic of a population, such as a mean or standard deviation.
- A statistic is a measurable characteristic of a sample, such as a mean or standard deviation.
- A sampling method is a method for selecting sample elements from a population.
- A random number is a number found totally by chance, with no predictable relationship to any other number.

A random number table is a list of numbers, comprises of the digits 0, 1, 2, 3, 4, 5, 6, 7, 8 and 9. Numbers in the list are arranged so that each digit has no predictable relationship to the digits that preceded it or to the digits that followed it. In short, the digits are arranged randomly. The numbers in a random number table are random numbers.

Simple Random Sampling

The simple random sampling refers to a sampling method that has the following properties:

- The population has N objects.

- The sample consists of n objects.
- All possible samples of n objects are equally likely to occur.

A major advantage of simple random sampling is that it allows researchers to use statistical methods to analyze sample results. For example, provided a simple random sample, researchers can use statistical methods to define a confidence interval around a sample mean. Statistical analysis is not appropriate when non-random sampling methods are used.

There are several methods to get a simple random sample. One way would be the lottery method. Each of the N population members is assigned a unique number. The numbers are placed in a bowl and totally mixed. Then, a blind-folded researcher selects n numbers. Population members having some selected numbers are included in the sample.

In this method, each member of the population has an equal chance of being selected as subject. The whole process of sampling is completed in a single step with each subject selected independently of the other members of the population.

There are several methods to proceed with simple random sampling. The most primitive and mechanical would be the lottery method. Every member of the population is assigned a unique number. Each number is placed in a bowl or a hat and mixed thoroughly. The blind-folded researcher then picks numbered tags from the hat. All the individuals bearing the numbers selected by the researcher are the subjects for the study. Another method would be to let a computer do a random selection from the population. For populations with a small number of members, it is advisable to use the initial method but if the population has several members, a computer-aided random selection is preferred.

Advantages of Simple Random Sampling

One of the important things about the simple random sampling is the ease of assembling the sample. It is also considered as a best way of selecting a sample from a provided population since every member is given equal opportunities of being selected.

Another key feature of simple random sampling is its representativeness of the population. If the sample is not representative of the population, the random variation is called as sampling error.

An unbiased random selection and a representative sample is necessary in drawing conclusions from the results of a study. Remember that one of the goals of research is to be able to create the conclusions pertaining to the population from the results obtained from a sample. Due to the representativeness of a sample got by simple random sampling, it is reasonable to create generalizations from the results of the sample back to the population.

Disadvantages of Simple Random Sampling

One of the most obvious limitations of simple random sampling method is its need of a complete list of all the members of the population. Please keep in mind that the list of the population must be complete and up-to-date. This list is usually not available for large populations. In cases as such, it is wiser to use other sampling techniques.

Assignable Causes

When the observed variations are not statistically found to be due to random effects, it clearly points to the existence of some assignable causes. In our detergent filling machine example, if we found that the variations are beyond random effects, then several other issues may contribute to the variations. If the machine requires some operator intervention, then differences in the practices between two operators could be one cause. Alternatively, the viscosity of the detergent powder could have changed due to some changes in the previous stage of production. A third possibility is malfunctioning of the filling machine itself due to some structural disturbances in the operating conditions of the machine or the operating environment. From a statistical point of view, assignable causes of variations impact the distribution of the process in more ways than one. One possibility is that the average would have shifted out of the desired level. Instead of operating with an average of 500 gins, the detergent filling machine might have shifted to some other mean (say 497 gins or 502 gms). The second possibility is that the variance in the process increases significantly, leading to a wider spread. When the spread is wider, large variations in the process will be considered to be within the chance category, when it is not the case in reality. It is also possible that the process is skewed to one side, thereby having a greater tendency to fill more than 500 gins or less than 500 gins depending on the direction of the skewing. In all the above cases, it is possible to trace the causes of the variations and take corrective actions. Once corrective actions are implemented, the system will return to normal operating conditions and variations will be only due to chance. A process having only chance variations is said to be in a state of control. Therefore, it is clear that when a process is under a state of control, the associated distribution of the process will not shift the average. Moreover, the spread of the process will also remain the same and the distribution will be symmetric. Before we proceed further in our understanding of the elements of process control, it is useful to know the terms associated with the techniques.

Nominal Value

The desired average for a process is known as the nominal value. In our detergent filling machine example, the nominal value for the weight of the filled sachets is 500 gms.

Specification Limits

Specification limits are nothing but the tolerances allowed in every process. These tolerances are arrived at after making a detailed design based on several considerations. The upper specification limit (USL) provides the upper limit for the measure under consideration, while the lower specification limit (LSL) provides the lower limit. In our detergent filling machine example, the design for the process would have specified that the process could have a fill weight ranging between 496 gins and 504 gins. In this case, USL 504 gins and LSL - 496 gms. The specification limits are nothing but the voice of the customer, pertaining to the process under consideration.

Process Average

How the process is operating is independent of what one desires the process to be. By actually sampling the process, one can establish the average at which the process is operating. This average is known as the center line (of the process).

Control Limits

As we already saw, the process is considered to be under control when there are only chance variations. The upper control limit (UCL) is defined by the value of three positive standard deviations from the process average. Similarly, the lower control limit (LCL) is defined by three negative standard deviations from the process average. It can be numerically expressed as follows:

- $UCL = \text{Process Average} + 3\sigma$
- $LCL = \text{Process Average} - 3\sigma$

Clearly, as actual observations fall within UCL and LCL, we conclude that there are only chance variations. The process average and the control limits represent the voice of the process. In our detergent example, if the process is centred at 500.5 gins and the standard deviation is 0.25, then UCL 501.25 gins and the LCL 499.75 gms.

4.2 The Principle of Shewhart Control Chart

The Shewhart control chart has a baseline and upper and lower limits, represented as dashed lines that are symmetric about the baseline. Measurements are plotted on the chart versus a time line. Measurements that are outside the limits are considered to be out of control.

The baseline for the control chart is the accepted value, an average of the historical check standard values. A minimum of 100 check standard values is needed to establish an accepted value.

The upper (UCL) and lower (LCL) control limits are:

- UCL = Accepted value + k*process standard deviation
- LCL = Accepted value - k*process standard deviation

Where the process standard deviation is the standard deviation calculated from the check standard database.

This method is an individual observations control chart. The previously described control charts depended on rational subsets, which use the standard deviations computed from the rational subsets to compute the control limits. For a measurement process, the subgroups would consist of short-term repetitions which can characterize the precision of the instrument but not the long-term variability of the process. In measurement science, the interest is in assessing individual measurements. Thus, the standard deviation over time is the appropriate measure of variability.

To find out tight control of the measurement process, set:

$$k = 2$$

In which case approximately 5% of the measurements from a process that is in control will produce out-of-control signals. This considers that there is a sufficiently large number of degrees of freedom (>100) for estimating the process standard deviation.

To flag only those measurements that is egregiously out of control, set:

$$k = 3$$

In which case approximately 1% of the measurements from an in-control process will create out-of-control signals.

Variable Control Charts

These are used when the quality characteristic can be measured and expressed in numbers.

- Length, Volume, Weight, Temperature, Time.

Examples of Variable Control Charts

- X and R.
- X and s.
- Delta.
- X and Moving Range.

2 charts are used together: R-chart ("range chart") and X bar chart ("average chart").

Both the process variability (measured by the R-chart) and the process average (measured by the X bar chart) should be in control before the process can be said to be in control.

Process variability should be in control before the X bar chart can be developed because a measure of process variability is required to determine the chart control limits.

R-Chart for Process Variability

$$UCL_R = D_4(R)$$

$$LCL_R = D_3(R)$$

Where the average of past R values and D3 and D4 is are the constant based on the sample size.

$\bar{X}$ Chart for Process Average

$$UCL_R = \bar{X}_{bar} + A_2(R)$$

$$LCL = \bar{X}_{bar} - A_2(R)$$

Where the X bar is the average of many past values and A_2 is a constant based on the sample size.

Mean () Chart

The $\bar{X}$ chart is based on the mean of a sample taken from the process under study. The sample contains 4 observations. To construct a mean chart we first require constructing the center line of the chart. This is got by calculating the means of each sample. The center line of the chart is then computed as the mean of all sample means, where shows the number of samples:

$$\bar{\bar{X}} = \frac{\bar{X}_1 + \bar{X}_2 + \ldots + \bar{X}_n}{n} = CL$$

To construct the upper and lower control limits of the chart, we use the following formulas:

Upper control limit (UCL)= $\bar{\bar{X}} + Z\sigma_{\bar{X}}$

Lower control limit (LCL) = $\bar{\bar{X}} - Z\sigma_{\bar{X}}$

Where $\bar{X}.$ = the average of the sample means.

Z = Standard normal variable (2 for 95.44% confidence, 3 for 99.74% confidence).

$\sigma_{\bar{X}}$ = Standard deviation of the distribution of sample means, determined as: $\sigma / \sqrt{n}$.

σ = Population (process) standard deviation.

Sample size (number of observations per sample).

Alternatively,

The control limits can be constructed by using the sample range as an estimate of the variability of the process. The control limits for the range, R (max-value – min-value in the sample) is provided as:

$$UCL = \bar{\bar{X}} + A_2\bar{R}$$

$$LCL = \bar{\bar{X}} - A_2\bar{R}$$

$\bar{R}$: =Average range of the samples, $\bar{R} = \frac{\sum_i^k = 1\bar{R}}{K}$.

A_2= Factor obtained from the table.

NB: A_2 is a factor that includes 3 standard deviations of ranges and is dependent on the sample size being considered.

Constructing the Range (R) Chart

The R-Chart is constructed in a manner same as the - Chart. The center line of the control chart is the average range and the upper and lower control limits are computed as follows:

$$CL = \bar{R} = \frac{\sum_i^k = 1\bar{R}}{K}$$

$$UCL = D_4\bar{R}$$

$$LCL = D_3\bar{R}$$

Whereas the values of D_3 and D_4 are got from the table. When $D_3 = 0 \forall n \leq 6$. $LCL_R = 0$. Actually, three signal limits often yield a negative lower control limit, which is recorded as zero. This gives that, with samples of six (6) or fewer, it will be not possible for value on the R-Chart to fall outside the lower limit. In effect, the R-Chart will not be capable of detecting reductions in the dispersion of the process output.

Limitations

- It should be able to measure the quality characteristics in numbers.
- It might be impractical and uneconomical.

E.g. manufacturing plant responsible of 100,000 dimensions.

Attribute Control Chats

The attribute Control Charts are used for product characteristics that can be evaluated with a discrete response (yes/no, pass/fail, good/bad, number defective).

Examples of Attribute Control Charts

- np chart
- p chart
- u chart
- c chart

The Shewhart control chart plots quality characteristics that can be measured and expressed numerically. We measure height, weight, position, thickness, etc. If we can't represent a particular quality characteristic numerically or if it is impractical to do so, we then usual resort to using a quality characteristic to sort or classify an item that is inspected into 1 of 2 "buckets".

An example of a general quality characteristic classification would be designating units as "conforming units" or "nonconforming units". Another quality characteristic rule will be sorting units into "non-defective" and "defective" categories. Quality characteristics of that kind are called attributes.

Note that there is a difference between "nonconforming to an engineering specification" and "defective" a nonconforming unit might function just fine and be, in fact, not defective at all, while a part can be "in spec" and not function as expected(i.e., be defective).

Examples of quality characteristics that are attributes are the number of failures in a production run, the proportion of malfunctioning wafers in a lot, the number of people eating in the cafeteria on a provided day, etc.

Control charts dealing with the number of defects or nonconformities are known as c charts (for count).

Control charts dealing with the proportion or fraction of defective product are known as p charts (for proportion).

There is another chart which handles defects per unit, called the u chart (for unit). This gives when we wish to work with the average number of nonconformities per unit of product.

For additional references, see Woodall (1997) which reviews papers showing examples of attribute control charting, including examples from semiconductor manufacturing such as those examining the spatial dependence of defects.

Constructing the P - Chart

The computation of the center line as well as the upper and lower control limits is same as the computation for other type of control charts. CL is computed as the average proportion defective in the population, $\overline{P}$:

$$\overline{P} = \frac{\text{Total number of Defective Items}}{\text{Total number of observation}}$$

$$UCL = \overline{P} + Z\sigma_p$$

$$LCL = \overline{P} - Z\sigma_p$$

Where Z =Standard normal variable.

$\overline{P}$ =The sample proportion defective.

σ_p= The standard deviation of the average proportion.

As with the other charts, Z is chosen to lie within three standard deviation. The sample standard deviation is computed as follows:

$$\sigma_p = \sqrt{\frac{\overline{P}(1-\overline{P})}{n}}$$

Where n is the sample size.

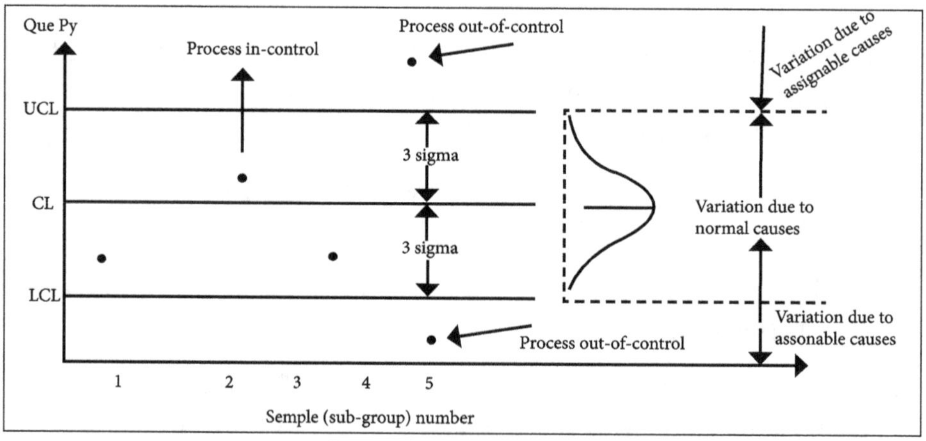

A specimen of a typical control chart.

Where,

UCL=Upper Control Limit

CL=Control Limit (center line)

LCL=Lower Control Limit.

Limitations

- In general are less costly when it comes to collecting data.
- It can plot multiple characteristics on one chart.
- But, loss of information vs. variable chart.

4.2.1 Constructions and Operation of X- Bar Chart

$\bar{x}$ Chart: Process Mean and Standard Deviation Known

To illustrate the construction of an $\bar{x}$ chart, let us consider the situation at KJW Packaging. This operates a production line where cartons of cereal are filled. Suppose KJW knows that when the process is operating correctly and hence the system is in control the mean filling weight is $\mu = 16.05$ ounces and the process standard deviation is a = 0.10 ounces. In addition, assume the filling weights are normally distributed. This distribution is shown in Figure.

The sampling distribution of $\bar{x}$ can be used to determine the expected variation in $\bar{x}$ values for a process that is in control. Let us first briefly review the properties of the sampling distribution of $\bar{x}$. First, recall that the expected value or mean of $\bar{x}$ is equal to μ, the mean filling weight when the production line is in control.

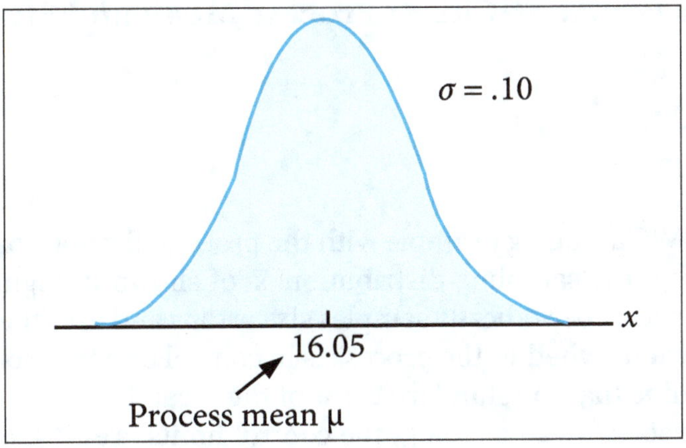

For samples of size n, the formula for the standard deviation of $\bar{x}$, called the standard error of the mean, is:

$$\sigma_{\bar{x}} = \frac{\sigma}{\sqrt{n}}$$

In addition, because the filling weights are normally distributed, the sampling distribution of $\bar{x}$ is normal for any sample size. Thus, the sampling distribution of $\bar{x}$ is a normal probability distribution with mean μ and standard deviation $\sigma_{\bar{x}}$. This distribution is shown in figure.

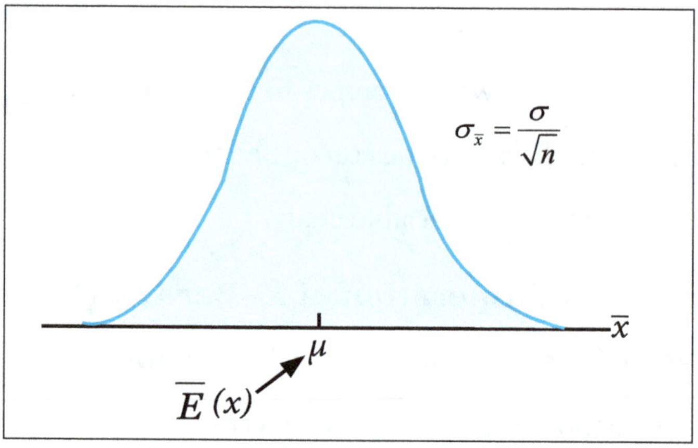

The sampling distribution of $\bar{x}$ is used to determine what values of $\bar{x}$ are reasonable if the process is in control. The general practice in quality control is to define as reasonable any value of $\bar{x}$ that is within 3 standard deviations above or below the mean value. Recall from the study of the normal probability distribution that approximately 99.7% of the values of a normally distributed random variable are within 3 standard deviations of its mean value. Thus, if a value of $\bar{x}$ is within the interval $\mu - 3\sigma_{\bar{x}}$ to $\mu + 3\sigma_{\bar{x}}$, we will assume that the process is in control. In summary, then, the control limits for an $\bar{x}$ chart are as follows.

Control Limits For An $\bar{X}$ Chart: Process Mean and Standard Deviation Known

$$\text{UCL} = \mu + 3\sigma_{\bar{x}}$$
$$\text{LCL} = \mu + 3\sigma_{\bar{x}}$$

Reconsider the KJW Packaging example with the process distribution of filling weights shown in Figure and the sampling distribution $\bar{x}$ of shown in Figure. Assume that a quality control inspector periodically samples six cartons and uses the sample mean filling weight to determine whether the process is in control or out of control. Using equation (1), we determine that the standard error of the mean is $\sigma_{\bar{x}} = \sigma / \sqrt{n} = .10 / \sqrt{6} = .04$. Thus, with the process mean at 16.05, the control limits are UCL = 16.05 + 3(.04) = 16.17 and LCL = 16.05-3(.04) = 15.93. Figure is the control chart with the results of 10 samples taken over a 10-hour period. For ease of reading, the sample numbers 1 through 10 are listed below the chart.

Note that the mean for the fifth sample in figure shows that the process is out of control. In other words, the fifth sample mean is below the LCL indicating that assignable causes of output variation are present and that under filling is occurring. As a result, corrective action was taken at this point to bring the process back into control. The fact that the remaining points on the $\bar{x}$ chart are within the upper and lower control limits indicates which the corrective action was successful.

4.2.2 R-Chart, P-Chart and C-Chart

R Charts

Let us now consider a range chart (R chart) that can be used to control the variability of a process. To develop the R chart, we need to think of the range of a sample as a random variable with its own mean and standard deviation. The average range provides an estimate of the mean of this random variable. Moreover, it will be shown that an estimate of the standard deviation of the range is:

$$\hat{\sigma}_R = d_3 \frac{\bar{R}}{d_2}$$

Where d_2 and d_3 are constants that depend on the sample size; values of d_2 and d_3 are also provided in Table. Thus, the UCL for the R chart is given by:

$$\bar{R} + 3\hat{\sigma}_R = \bar{R}\left(1 + 3\frac{d_3}{d_2}\right)$$

And the LCL is:

$$\bar{R} - 3\hat{\sigma}_R = \bar{R}\left(1 - 3\frac{d_3}{d_2}\right)$$

If we let:

$$D_4 = 1 + 3\frac{d_3}{d_2}$$

$$D_3 = 1 - 3\frac{d_3}{d_2}$$

Values for D_3 and D_4 are also provided in Table. Note that for n = 5, D_3= 0 and D_4 = 2.114. Thus, with R =.0253, the control limits are:

- $UCL = 0.0253(2.114) = 0\ .0534$
- $LCL = 0.0253(0) = 0$

Figure shows the R chart for the Jensen Computer Supplies problem. We used the data in Table 2 and Minitab's control chart routine to construct the chart. The center line is shown at the overall mean of the 20 sample ranges, $\bar{R} = 0.02527$. The UCL is .05344 or 3 sigma limits (3.0SL) above $\bar{R}$. The LCL is 0.0 or 3 sigma limits below $\bar{R}$. The R chart shows the 20 sample ranges plotted over time. Because all 20 sample ranges are within the control limits, we ensure that the process was in control throughout the sampling period. This chart can now be used to monitor the process variability on an ongoing basis.

P Chart

Let us consider the case in which the output quality is measured by either non defective or defective items. The decision to continue or to adjust the production process will be based on $\bar{p}$, the proportion of defective items found in a sample. The control chart employed for proportion-defective data is known as a p chart.

To show the construction of a p chart, consider the use of automated mail-sorting machines in a post office. These automated machines scan the zip codes on letters and divert every letter to its proper carrier route. Even when a machine is operating properly, some letters are diverted to incorrect routes. Assume that when a machine is operating correctly or in a state of control, 3 percent of the letters are incorrectly diverted. Thus p, the proportion of letters incorrectly diverted when the process is in control, is .03.

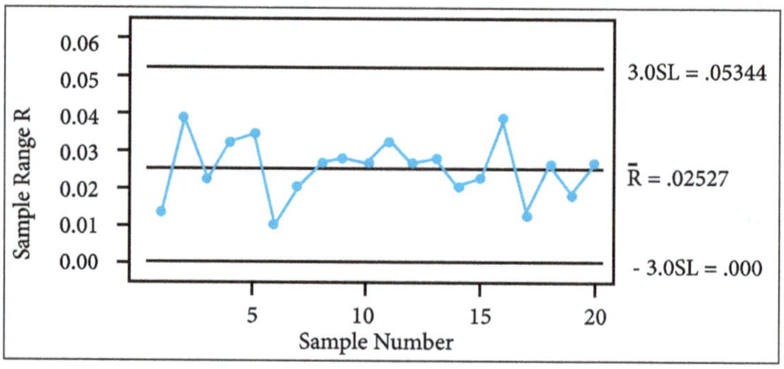

The sampling distribution of $\bar{p}$ will be used to find out the variation that can be expected in $\bar{p}$ values for a process that is in control. The expected value or mean of $\bar{p}$ is p, the proportion defective when the process is in control. With samples of size n, the formula for the standard deviation of $\bar{p}$, called the standard error of the proportion, is:

$$\sigma_{\bar{p}} = \sqrt{\frac{p(1-p)}{n}}$$

The sampling distribution of $\bar{p}$ can be approximated by a normal probability distribution whenever the sample size is large. With $\bar{p}$, the sample size can be taken as large when-ever the following 2 conditions are satisfied.

$$np \geq 5$$
$$n(1-p) \geq 5$$

In summary, whenever the sample size is large, the sampling distribution of $\bar{p}$ can be approximated by the normal probability distribution with mean p and standard deviation $\sigma_{\bar{p}}$.This distribution is represented in Figure.

To establish control limits for a p chart, we follow the same procedure we employed to establish the control limits for a $\bar{x}$ chart. That is, the limits for the control chart are set at 3 standard deviations or standard errors, above and below the proportion defective when the process is in control. Thus, we have the following control limits.

Control Limits for A P Chart

$$UCL = p + 3\sigma_{\bar{p}}$$

$$LCL = p - 3\sigma_{\bar{p}}$$

Steps in Making the Xbar and R Chart

- Collect the data. It is best to have at least hundred samples.
- Divide the data into sub groups, it is recommended the subgroups be of 4 or 5 data points each. The number of samples is shown by the letter "n" and the number of subgroups is shown by the letter "k". The data should be divided into subgroups in keeping with the following conditions:
  - The data obtained should be from the same grouping of products produced.
  - A sub group must not include data from a different lot or different process.
- Record the data on a data sheet. Design the sheet so that it is simple to compute the values of X bar and R for each sub group.
- Find the mean value (X bar). Use the following formula for each subgroup:

$$\bar{X} = \frac{x1 + x2 + x3 + x4 + x5}{N} = \frac{65.0}{5} = 13.0$$

- Find the range, R. Use the following formula for each subgroup.

$$R = X(\text{largest value}) - X\ (\text{smallest value})$$

X Bar & R Control Chart: Class Exercise

Sub Group	6:00	10:00	14:00	18:00	22:00	X bar	R
1	14.0	12.6	13.2	13.1	12.1	13.00	1.9
2	13.2	13.3	12.7	13.4	12.1	12.94	1.3
3	13.5	12.8	13.0	12.8	12.4	12.90	1.1
4	13.9	12.4	13.3	13.1	13.2	13.18	1.5
5	13.0	13.0	12.1	12.2	13.3	12.72	1.2
6	13.7	12.0	12.5	12.4	12.4	12.60	1.7
7	13.9	12.1	12.7	13.4	13.0	13.02	1.8
8	13.4	13.6	13.0	12.4	13.5	13.18	1.2

9	14.4	12.4	12.2	12.4	12.5	12.78	2.2
10	13.3	12.4	12.6	12.9	12.8	12.80	0.9
11	13.3	12.8	13.0	13.0	13.1	13.04	0.5
12	13.6	12.5	13.3	13.5	12.8	13.14	1.1
13	13.4	13.3	12.0	13.0	13.1	12.96	1.4
14	13.9	13.1	13.5	12.6	12.8	13.18	1.3
15	14.2	12.7	12.9	12.9	12.5	13.04	1.7
16	13.6	12.6	12.4	12.5	12.2	12.66	1.4
17	14.0	13.2	12.4	13.0	13.0	13.12	1.6
18	13.1	12.9	13.5	12.3	12.8	12.92	1.2
19	14.6	13.7	13.4	12.2	12.5	13.28	2.4
20	13.9	13.0	13.0	13.2	12.6	13.14	1.3
21	13.3	12.7	12.6	12.8	12.7	12.82	0.7
22	13.9	12.4	12.7	12.4	12.8	12.84	1.5
23	13.2	12.3	12.6	13.1	12.7	12.78	0.9
24	13.2	12.8	12.8	12.3	12.6	12.74	0.9
25	13.3	12.8	13.0	12.3	12.2	12.72	1.1
						$\sum \bar{x} = 323.50$	$\sum R = 33.8$
						$\bar{\bar{x}} = 12.940$	$\bar{R} = 1.35$

n	A2	D4	D3
2	1.88	3.267	---
3	1.023	2.575	---
4	0.729	2.282	---
5	0.577	2.115	---
6	0.483	2.004	---
7	0.419	1.924	0.076

- Let us calculate the remaining Xbar and R values from the above tabular column.

  Let us find the overall mean or X double bar $\bar{x}$.

- Total the mean values of X-bar, for each subgroup and divided by the number of subgroups (k).

$$\frac{\bar{x} = 13.0 + 12.94 + 12.90 ... + 12.72}{25} = \frac{323.50}{25} = 12.94$$

- Calculate the average value of the range (R). Total R for all the groups and divide by the number of subgroups (k).

$$\bar{R} = \frac{R1 + R2 + R3 ... Rk}{k}$$

$$\bar{R} = \frac{1.9+1.3+1.1\ ...\ +1.1}{25} = \frac{33.8}{25} = 1.35$$

C Chart

Generally, C charts are used to chart the total number of defects in a sample when the sample size is constant. We can inspect for one type of defect such as dead pixels. We can also inspect for several defects together such as dead pixels, stuck pixels, scratches, and blurry spots. An LCD screen might have 2 or 3 dead pixels, yet still be acceptable.

Example of a C Chart

An LCD manufacturer wants to monitor defects on 17-inch LCD screens. Technicians record the number of dead pixels for each subgroup of 10 screens per hour. They use a C chart to monitor the number of dead pixels.

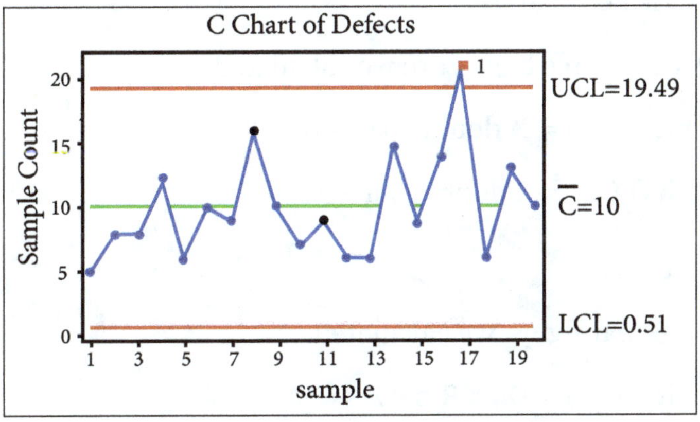

On average, technicians find 10 dead pixels in each sample. Sample 17 is out of control. The technicians should try to identify any special causes that might have contributed to the unusually high number of dead pixels.

Problems

1. Let us compute the average value of the range (R). Total R for all the groups and divide by the number of subgroups (k).

Solution:

$$\bar{R} = \frac{R1+R2+R3...Rk}{k}$$

$$\bar{R} = \frac{1.9+1.3+1.1\ ...\ +1.1}{25} = \frac{33.8}{25} = 1.35$$

Let us compute the Control Limit Lines. Use the following formulas for X-bar and R

Control Charts. The coefficients for determining the control lines are A2, D4 and D3 are located on the bottom of the Work Sheet we are presently using and presented here:

n	A2	D4	D3
2	1.88	3.267	---
3	1.023	2.575	---
4	0.729	2.282	---
5	0.577	2.115	---
6	0.483	2.004	---
7	0.419	1.924	0.076
8	0.373	1.864	0.136
9	0.337	1.816	0.184
10	0.308	1.777	0.223

Xbar Control Chart

Central Line (CL) = X double bar figure we calculated.

Upper Control Limit (UCL) = X double bar + A2* R bar.

Lower Control Limit (LCL) = X double bar - A2* R bar.

R Control Chart

Central Line (CL) = R bar figure we calculated.

Upper Control Limit (UCL) = D4 * R bar.

Lower Control Limit (LCL) = D3 * R bar.

Details are as follows:

X Control Chart CL = X double bar = 12.94.

UCL = 12.94 + .577 * 1.35 = 13.719, Note that we are using five subgroups, so on the chart n = 5 and under the A2 column, 5 = 0.577. 1.35 is the figure we determined for R bar.

LCL = 12.94 - .577 * 1.35 = 12.161

R Control Chart CL = R bar = 1.35

UCL = 2.115 * 1.35 = 2.86 Note that we are using 5 subgroups, so on the chart n = 5 and under the D4 column, 5 = 2.115.

LCL = since our subgroups equal 5, if we look under the D3 column, there is no calculation coefficient to apply, thus there is no LCL.

Construct the Control Chart. Using graph paper or Control Chart paper set the index so that the upper and lower control limits will be splitted by 20 to 30 mm (units). Let us draw in the Control lines CL, UCL and the LCL and label them with their appropriate numerical values. It is recommended that we use a blue or black line for the CL and a red line for the UCL and LCL. The central line is a solid line. The Upper and Lower control limits are usually plotted as broken lines.

Plot the Xbar and R values as computed for every subgroup. For the Xbar values, use a dot (.) and for the R values, use an (x). Circle any points that lie outside the control limit lines so that we can distinguish them from the others. The plotted points should be about 2 to 5 mm apart. Below figure represents how our X-bar chart looks like when plotted.

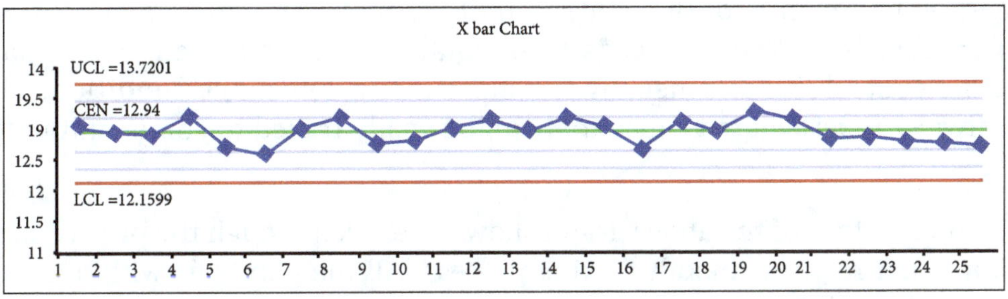

Below figure represents how our R-bar chart looks like when plotted.

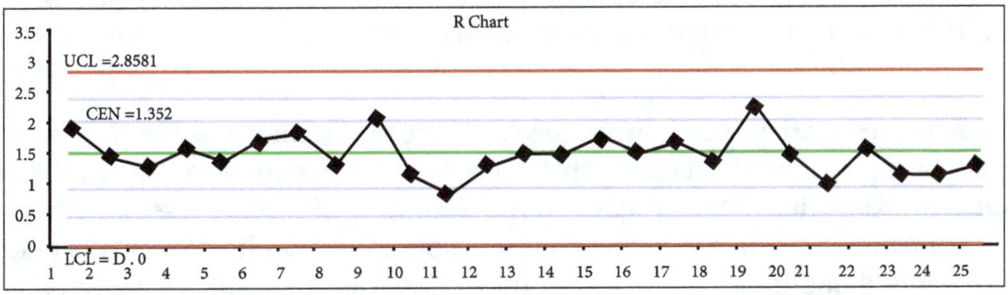

Analysis of the Control Chart

Now that we know the how to make a control chart, it is even more necessary to understand how to interpret them and realize when there is a problem. All processes have some type of variation and this process variation can be partitioned into 2 main components. Initially, there is natural process variation, frequently known as "common cause" or system variation. These are common variations caused by machines, material and the natural flow of the process. Secondly is special cause variation, generally caused by some problem or extraordinary occurrence in the system. It is our job to work at trying to eliminate or minimize both of these kinds of variation. Below is an example of a few different process variations and how to recognize a potential problem.

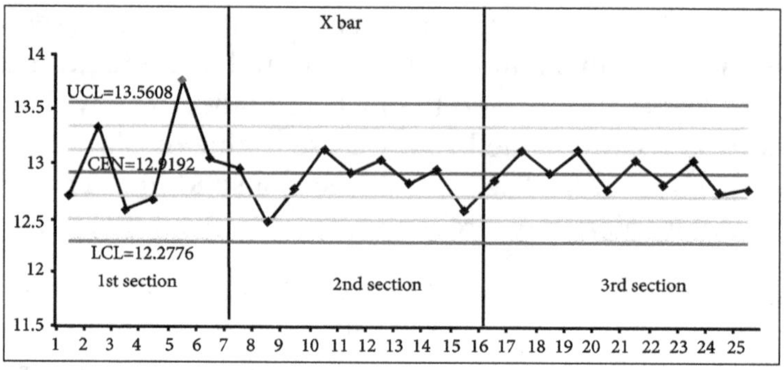

In the above chart, there are 3 divided sections. The first section of the graph termed "out of statistical control" for many reasons. Notice the inconsistent plot points and that one point is outside of the control limits. This means that a source of special cause variation is present, it needs to be analyzed and resolved. Having a point outside the control limits is usually the most easily detectable condition. There is almost always an associated cause that can be easily traced to some malfunction in the process.

In the second section of the above graph shows like, even though the process is now in control, it is not really a smooth flowing process. All the points lie within the control limits and thus exhibits only common cause variations.

In the third section, we will notice that the trending is more predictable and smoother flowing. It is in this section that there is evidence of process improvement and the variation has been reduced.

Therefore, to summarize, eliminating special cause variation keeps the process in control, process improvement decreases the process variation and moves the control limits in toward the centerline of the process. At the beginning of this process run, it was in require of adjustment as the product output was sporadic. An adjustment was made and while the plotted points were now within the boundaries, it is still not centered around the process specification. Atlast, the process was tweaked a little more and in the third section, the process seems to center around the CL.

There are a few more terms listed below that we require to become familiar with when analyzing a X-bar Chart and the process:

TREND - If there is a continued rise or fall in a series of points (like an upward or downward slant), it is considered a "trend" and generally denotes a process is drifting out of control. This usually needs a machine adjustment.

RUN - When many plotted points line up consecutively on one side of a Central Line (CL), whether it is located above or below the CL, it is called a "run". If there are 7 points in a row on one side of the CL, there is an abnormality in the process and it needs an adjustment.

HUGGING - When the points on the control chart seem to stick close to the center line or to a control limit line, it is called "hugging of the control line". This generally indicates that a different type of data or data from different factors (or lines) have been mixed into the sub groupings. To determine if we are experiencing "hugging" of the control line, perform the following exercise. Draw a line equal distance between the centerline and the upper control limit. Then draw another line equal distance between the center line and the lower control limit. If the points remain inside of these new lines, there is an abnormality and the process requires closer analysis.

PERIODICITY - If the plotted points show the same pattern of change over equal intervals, it is called "periodicity". It looks much like a uniform roller coaster of the same size ups and downs around the centerline. This process must be watched closely as something is causing a defined uniform drift to both sides of the centerline.

Problems

1. A quality control inspector at a soft drink company has taken 25 samples each with 4 observations on the volume of bottle filled. Standard deviation of bottling operation is 0.14 oz.

Develop Control Charts for the inspection of the bottling operation.

Inspection Data

Sample Number	Observation (bottle volume in ounces)				Average	Range
	1	2	3	4	$\overline{X}$	R
	15.85	16.02	15.83	15.93	15.91	0.19
	16.12	16.00	15.85	16.01	15.99	0.27
	16.00	15.91	15.94	15.83	15.92	0.17
	16.20	15.85	15.74	15.93	15.93	0.46
	15.74	15.86	16.21	16.10	15.98	0.47
	15.94	16.01	16.14	16.03	16.03	0.20
	15.75	16.21	16.01	15.86	15.96	0.46
	15.82	15.94	16.02	15.94	15.93	0.20
	16.04	15.98	15.83	15.98	15.96	0.21
	15.64	15.86	15.94	15.89	15.83	0.30
	16.11	16.00	16.01	15.82	15.99	0.29
	15.72	15.85	16.12	16.15	15.96	0.43
	15.85	15.76	15.74	15.98	15.83	0.24
	15.73	15.84	15.96	16.10	15.91	0.37
	16.20	16.01	16.10	15.89	16.05	0.31
	16.12	16.08	15.83	15.94	15.99	0.29

	16.01	15.93	15.81	15.68	15.86	0.33
	15.78	16.04	16.11	16.12	16.01	0.34
	15.84	15.92	16.05	16.12	15.98	0.28
	15.92	16.09	16.12	15.93	16.02	0.20
	16.11	16.02	16.00	15.88	16.00	0.23
	15.98	15.82	15.89	15.89	15.90	0.16
	16.05	15.73	15.73	15.93	15.86	0.32
	16.01	16.01	15.89	15.86	15.94	0.15
	16.08	15.78	15.92	15.98	15.94	0.30
Total					398.75	7.17

Solution:

X – Bar and R Control Chart Limits

X – Bar Chart

$CL\ \bar{\bar{x}} = 15.95$

$UCL = 16.16$

$LCL = 15.74$

R – Chart

$CL\ \bar{R} = 0.29$

$UCL = 0.6612$

$LCL = 0$

X Bar Chart

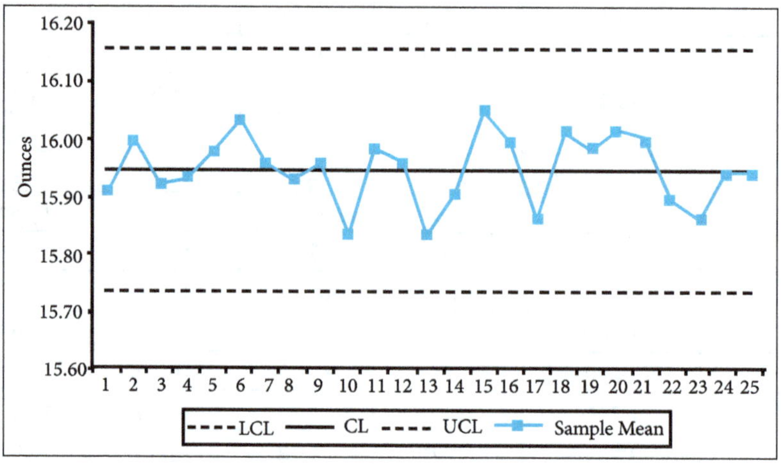

R Chart

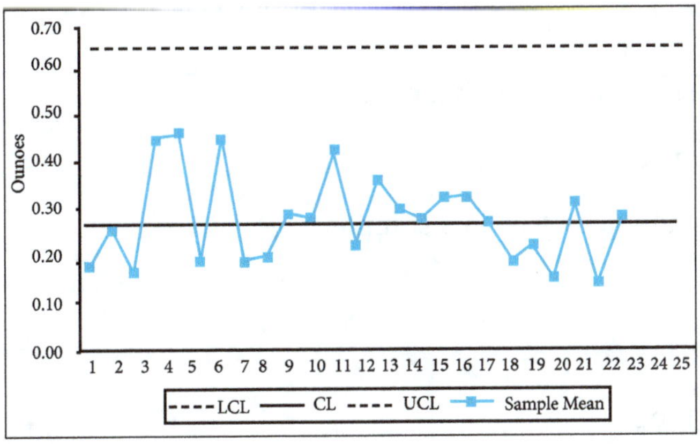

Interpreting X Bar and R Charts

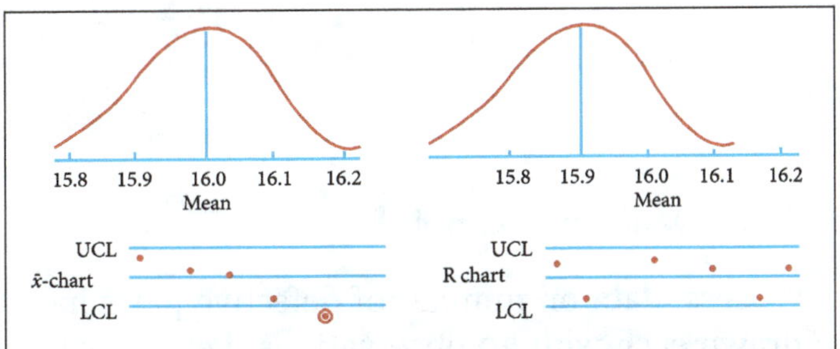

(a) Shift in mean detected by $\bar{x}$-chart but not by R-chart

Interpreting X Bar and R Charts

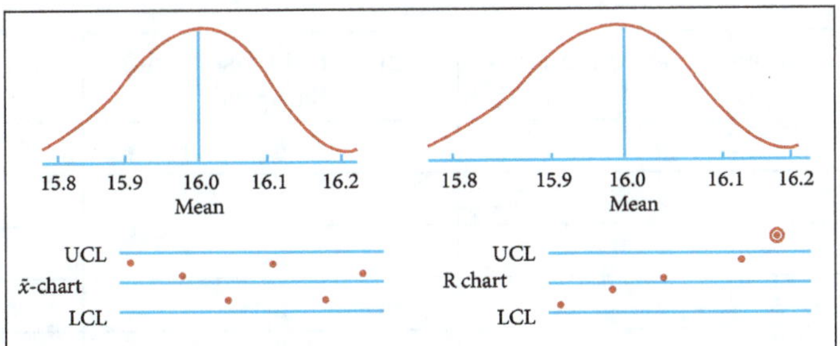

(b) Shift in dispersion detected by R-chart but not by $\bar{x}$-chart

Inspection by Attributes

Measures the Quality characteristics for parts with Attributes which are Discrete in nature.

- The number of Defects in a sample, Broken cookies in a pack, Rotten apples in a box.
- Both the Sample size and number of defects are Countable.

P Control Chart

Chart constructed in terms of:

p = Average Proportion Defective in a sample.

P Control Chart limits.

$\overline{p}$ = average proportion defective in a sample.

Chart Limits

Center line: $\overline{p}$

$$UCL = \overline{p} + 3 * \sigma_p$$

$$LCL = \overline{p} - 3 * \sigma_p$$

$$\sigma_p = \sqrt{\frac{\overline{p}(1-\overline{p})}{n}}$$

Where n=number of observation in a sample.

2. In a Production Process, data on number of defective parts has been collected. 20 samples have been drawn each with 20 observations. Let us construct P chart for the inspection process.

Inspection Data – Number of Defects

Sample Number	Number of Defective Tires	Number of Observation Sampled	Fraction Defective
	3	20	.15
	2	20	.10
	1	20	.05
	2	20	.10
	1	20	.05
	3	20	.15
	3	20	.15
	2	20	.10
	1	20	.05
	2	20	.10
	3	20	.15
	2	20	.10
	2	20	.10

	1	20	.05
	1	20	.05
	2	20	.10
	4	20	.20
	3	20	.15
	1	20	.05
	1	20	.05
Total	40	400	

Solution:

P Chart Control Limits

Sample size (k) : 20

Observations per sample (n) : 20

Total Observations: 400

Total number of defects observed: 40

Fraction defective $\overline{p}$ = 0.1

$$\sigma_p = 0.067$$

Chart Limits

$$UCL = 0.301$$

$$LCL = -0.1 \; --> 0.0$$

P Chart

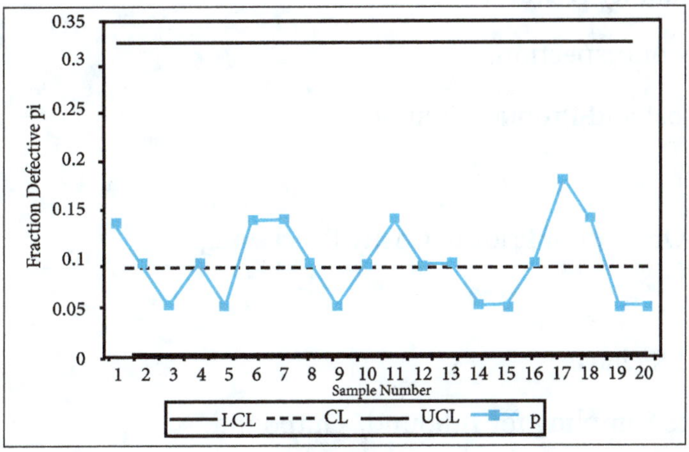

Process Capability

Evaluating ability of Process to meet Product Tolerances.

Process Capability Index-C_p

$$C_p = \frac{USL - LSL}{6\sigma}$$

Upper /Lower Specification limits

Process Capability C_p

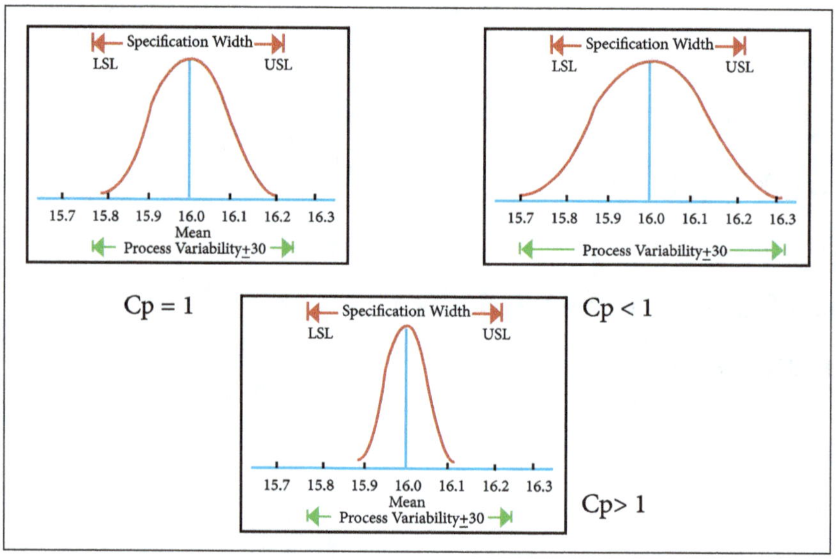

SQC – Implications for Managers

How much and How often to inspect?

- Process Stability History.
- Frequency of Inspection.
- Product Cost and Product Volume.

Where to Inspect?

- Inbound/Outbound/Prior to Costly Processing.

SQC Tools used:

- X-Bar and R charts.
- Acceptance Sampling for Inbound/ Outbound.

C-Chart

Sample No.	Number of Defects
1	5
2	4
3	6
4	5
5	4
6	5
7	5
8	4
9	5
10	6
11	4
12	6
13	6
14	8
15	7
16	6
17	4
18	7
19	6
20	6

Data for defects on TV set from 20 samples (sample size = 10) are shown in the table below:

$$CL = \frac{\text{sum of defects}}{\text{number of samples}} = \frac{109}{20} = 5.45$$

$$UCL = \bar{c} + 3\sqrt{\bar{c}} = 5.45 + 3\sqrt{5.45} = 12.45$$

$$LCL = \bar{c} - 3\sqrt{\bar{c}} = 5.45 - 3\sqrt{5.45} = -1.55 = 0$$

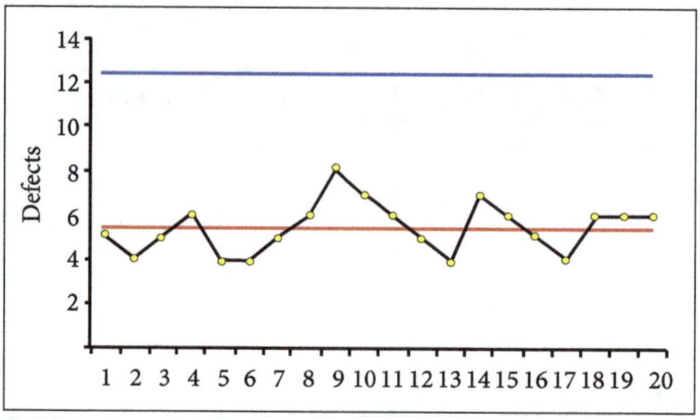

C- Chart

None of the sample is out of the LCL and UCL. But the chart shows cyclic trend.

P-Chart

Sample No.	No. of Defective CDs = x	Proportion Defective = x/sample size
1	4	.04
2	3	.03
3	3	.03
4	5	.05
5	6	.06
6	5	.05
7	2	.02
8	3	.03
9	5	.05
10	6	.06
11	6	.06
12	5	.05
13	4	.04
14	5	.05
15	4	.04
16	7	.07

17	6	..06
18	8	.08
19	6	.06
20	8	.08

$$CL = \frac{\text{sum of defectives}}{\text{sum of all samples}} = \frac{101}{2000} = 0.051$$

$$UCL = \bar{p} + 3\sqrt{\frac{\bar{p}(1-\bar{p})}{n}} = 0.051 + 3\sqrt{\frac{.051(1-.051)}{100}} = .066$$

$$LCL = \bar{p} - 3\sqrt{\frac{\bar{p}(1-\bar{p})}{n}} = 0.051 - 3\sqrt{\frac{.051(1-.051)}{100}} = .036$$

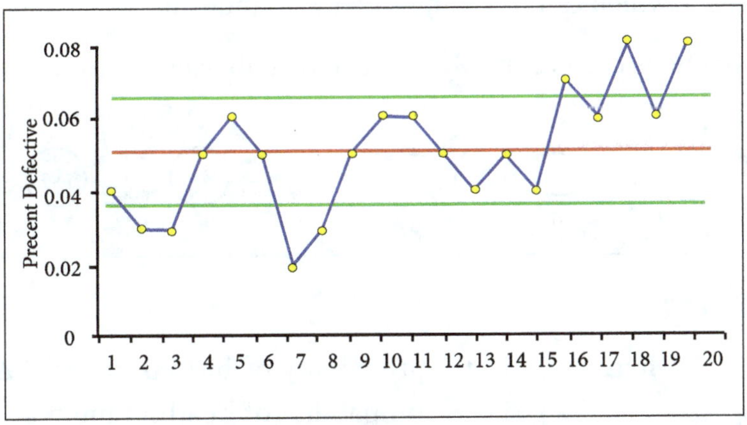

P- Chart

Sample data at S.N 16, 18 and 20 are above the UCL. Efforts should be made to determine the special causes and revised limits are advised to calculate after deleting these data. There is important observation that is clearly visible from the data points that there is an increasing trend in the average proportion defectives beyond sample number15 also, data show cyclic pattern. Process appears to be out of control and also there is strong evidence that data are not from independent source.

5

Queuing Theory

5.1 Pure Birth and Death Process

General Description

A birth-death (BD process) process refers to a Markov process with:

- A discrete state space,
- The states of which can be enumerated with index i= 0,1,2,... such that and
- State transitions can occur only between neighboring states, $i \rightarrow i+1$ or $i \rightarrow i-1$.

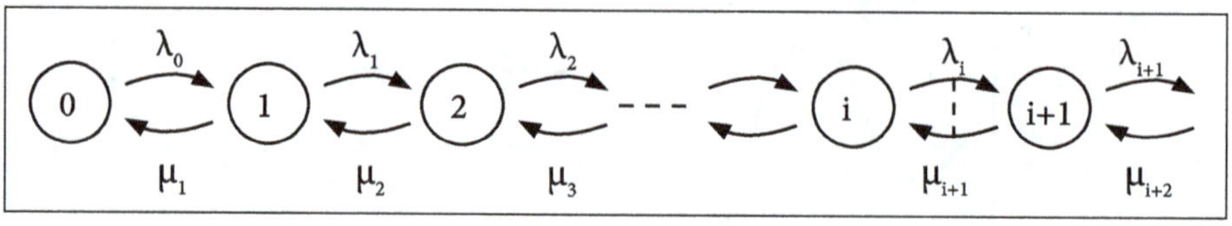

Transition Rates

$$q_{i,j} = \begin{cases} \lambda_i & \text{when} \quad j=i+1 \\ \mu_i & \text{if} \quad j=i-1 \\ 0 & \text{otherwise} \end{cases} \quad \begin{array}{l} \text{probability of birth in interval } \Delta t \text{ is } \lambda_i \, \Delta t \\ \text{probability of death in interval } \Delta t \text{ is } \mu_i \, \Delta t \\ \text{when the system is in state } i \end{array}$$

The Equilibrium Probabilities of a BD Process

We use the method of a cut = global balance condition applied on the set of states 0, 1, . . ., k. In equilibrium the probability flows across the cut are balanced (net flow =0):

$$\lambda_k \pi_k = \mu_{k+1} \pi_{k+1}, \quad \lambda_k \pi_k = \mu_{k+1} \pi_{k+1}$$

We obtain the recursion:

$$\pi_{k+1} = \frac{\lambda_k}{\mu_{k+1}} \pi_k$$

By means of the recursion, all the state probabilities can be expressed in terms of that of the state 0, π_0:

$$\pi_k = \frac{\lambda_{k-1}\lambda_{k-2}\ldots\lambda_0}{\mu_k\mu_{k-1}\ldots\mu_1}\pi_0 = \prod_{i=0}^{k-1}\frac{\lambda_i}{\mu_i+1}\pi_0$$

The probability πo is determined by the normalization condition πo:

$$\pi_o = \frac{1}{1+\frac{\lambda_o}{\mu_1}+\frac{\lambda_o\lambda_1}{\mu_1\mu_2}+\ldots} = \frac{1}{1+\sum_{k=1}^{\infty}\prod_{i=0}^{k-1}\frac{\lambda_i}{\mu_i+1}}$$

Problems

1. Let us define a Birth process.

Solution:

A Markov process $\{X(t)\}$ with state space $S=\{1,2.....\}$ such that

$$P\left[X(t+s_t)=k/X(t)=j\right]=\begin{cases}\lambda_k S_t, k=j+1, j\geq 1\\ 1-\lambda, s, k=j\geq 1\\ 0, \text{otherwise}\end{cases}$$ is called a birth process where , ,...... 1 2 λ

λ are the constant.

2. Let us define pure death process.

Solution:

$$\begin{cases}\lambda_i = 0\\ \mu_i = i\mu\end{cases} \quad i=0,1,2,.. \quad \pi_i(0)=\begin{cases}1 & i=n\\ 0 & i\neq n\end{cases}$$

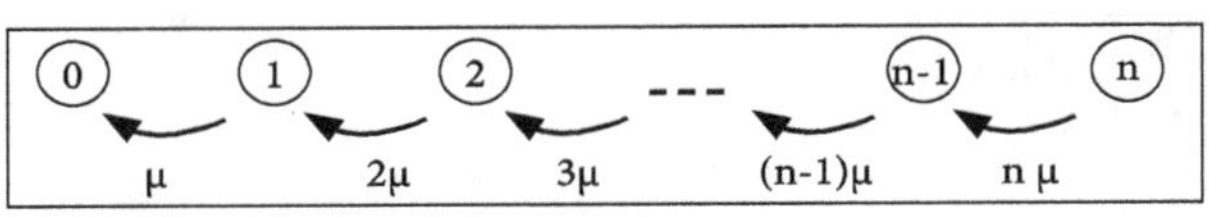

$$\begin{cases}\frac{d}{dt}\pi_n(t)=-n\mu\pi_n(t) & \Rightarrow \pi_n(t)=e^{-n\mu t}\\ \frac{d}{dt}\pi_i(t)=(i+1)\mu\pi_{i+1}(t)-i\mu\pi_i(t) & i=0,1,\ldots,n-1\end{cases}$$

$$\frac{d}{dt}\left(e^{i\mu t}\pi_i(t)\right)=(i+1)\mu\pi_{i+1}(t)e^{i\mu t} \quad \Rightarrow \pi_i(t)=(i+1)e^{-i\mu t}\mu\int_0^t \pi_{i+1}(t')e^{i\mu t}dt'$$

$$\pi_{n-1}(t)=ne^{-(n-1)\mu t}\,\mu\int_0^t \underbrace{e^{-n\mu t'}e^{(n-1)\mu t'}}_{e^{-\mu t'}}dt' = ne^{-(n-1)\mu t}\left(1-e^{-\mu t}\right)$$

Recursively $\pi_i(t)=\binom{n}{i}\left(e^{-\mu t}\right)^i\left(1-e^{-\mu t}\right)^{n-i}$. Binomial distribution: the survival probability at time t is e −μt independent of others.

3. Let us discuss about the steady state or long run probabilities for population size of a birth death process.

Solution:

Steady state probabilities of a birth-death process:

Transition occur only to the neighbouring state.

λ_k - birth rate when population size is K.

μ_k - death rate when population size is K(K * 0).

$P_i(t)$ = P (Population size is i at time t)

$$\frac{P_o(t+\Delta t)-P_o(t)}{\Delta t}=-\lambda_o P_o(t)+\mu \text{sub1 } P_1(t)$$

$$\frac{dP_o(t)}{dt}=-\lambda_o P_o(t)+\mu_1 P_1(t)$$

$$\frac{dP_i(t)}{dt}=-(\lambda_i+\mu_i)\,P_i(t)+\lambda_i-1\ P_i-1(t)\mu_i+1P_i+1(t)\ 1>0$$

In long run $\lambda_o P_o=\mu_1 P_1$

$$(\lambda_i+\mu_i)P_i=\lambda_{i-1}\ P_{i-1}+\mu_{i+1}P_{i+1}$$

Solving:

$$P_i=\frac{\lambda_o\,\lambda_i,\ \ldots,\ \lambda_i-1}{\mu_1\ \mu_2,\ \ldots,\ \mu_i}P_o$$

$$P_o=\left[\frac{1}{1}+\sum_{i=1}^{\infty}\prod_{i=0}^{i-1}\frac{\lambda_j}{\mu_j+1}\right]$$

5.2 M/M/1

The Utilization Difference Between a Multiple Server, Single Queue and a Multiple Server, Multiple Queue System

If we have a system with 4 processors where the arrival rate (λ) is 0.2 tasks per time unit and each processor has a service rate (μ) of 0.2 tasks per time unit.

Now we can have 2 different types of systems. System A has a single queue and 4 processors while the system B, 4 queues for each processor. The following diagrams depict the systems.

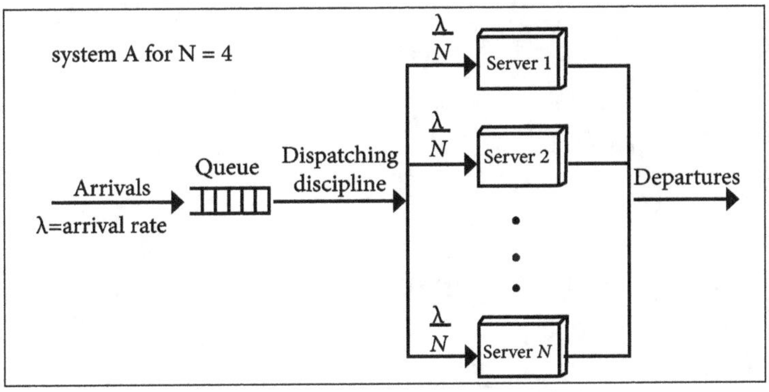

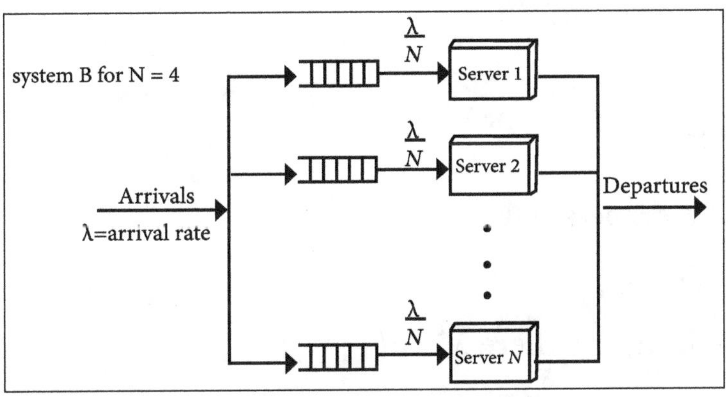

In which system the utilization of each server is better.

In system B, the average number of tasks in each queue and the average waiting time in each queue.

For the system A the utilization of each server will be $\lambda\mu m = 0.24 * 0.2 = 0.25$.

The systems B may be divide the arrival rate (λ) by 4 and treat each queue-server pair uniquely, then the utilization will be the same. So we will have no difference in the utilization. If we determine the utilization of each server, then since we know the arrival rate, finding the average waiting time and the average number of tasks in each queue will be very easy, however we do not know whether it's correct to say that the utilization is the same in both systems.

Single Server Finite Queue Model

Model (IV): (M/M/I): (N/FCFS)

$$P_o = \frac{1-p}{1-p}, \text{Where } p \neq 1,\ \rho = \lambda/\mu < 1$$

$$P_n = \begin{cases} p^n, & \dfrac{(1-\rho)}{} \\ & 0 \le n \le N, \rho \ne 1 (\lambda \ne \mu) \\ \dfrac{1}{N+1} & 1-\rho^{N+1} \\ & \rho = 1 (\lambda = \mu) \end{cases}$$

$$Ls = \frac{\lambda}{\mu-\lambda} - \frac{(N+1)(\lambda/\mu)^{N+1}}{1-(\lambda/\mu)^{N+1}}, \text{if } \lambda \ne \mu$$

(1) Average no of customers in the system:

$$L_s = \frac{\lambda}{\mu-\lambda} - \frac{(N+1)(\lambda/\mu)^{N+1}}{1-(\lambda/\mu)^{N+1}}, \text{ if } \lambda \ne \mu$$

$$\text{and } L = \sum_{n\ o} \frac{n}{K\ 1} = \frac{K}{2}, \quad \text{if } \lambda = \mu$$

(2) Average no of customers in the queue:

$$L_q = L_s - (1 - p_o)$$

$$L_q = L_s - \lambda/\mu, \quad \text{where } \lambda' = \mu(1 - p_o)$$

(3) Average waiting times in the system and in the queue:

$$W_s = L_s/\lambda' \ \& \ W_q = L_q/\lambda', \quad \lambda' = \mu(1-\rho_o)$$

Problems

1. Consider a single server queuing system with Poisson input, exponential service times, suppose the mean arrival rate is 3 calling units per hour, the expected service time is 0.25 hours and the maximum permissible no calling units in the system is two. Find the steady state probability distribution of the no of calling units in the system and the expected no of calling units in the system.

Solution:

$\lambda = 3$ units per hour

$\lambda = 4$ units per hour & $N = 2$

The traffic intensity $\rho = \lambda/\mu = 0.75$

$$P_n = \frac{(1-\rho)\rho^n}{1-\rho^{N+1}}, \rho \neq 1$$

$$=(0.43)(0.75)^n$$

$$P_0 = \frac{1-\rho}{1-\rho^{N+1}}, = \frac{1-0.75}{(1-0.75)^{2+1}} = 0.431$$

$$L_s = \sum_{n=1}^{W} n P_n = 0.81$$

2. An airport has a single runway. Airplanes have been found to arrive at the rate of 15 per hour. It is estimated that each landing takes 3 minutes. Assuming a Poisson process for arrivals and an exponential distribution for landing times, use an M/M/1 model to determine the following performance measures.

Solution:

a. Runway utilization:

Arrival rate = 15/hr (λ)

Service rate = 60/3/hr = 20/ hr (μ)

b. Expected number of airplanes waiting to land:

$$L_q = \frac{\rho^2}{1-\rho} = \frac{(.75)^2}{.25} = 2.25$$

c. Expected waiting time:

$$E(W) = \frac{}{\mu(\mu-\lambda)} = \frac{15}{20(20-15)} = \frac{3}{20} \text{hr} = 9 \text{mins}$$

d. Probability that the waiting will be more than 5 mins? 10 mins? No waiting?

$$P(\text{no waiting}) = P(T_q = 0) = 1-\rho = .25$$

$$P(T_q > t) = \rho e^{-\mu(1-\rho)t}$$

$$P(T_q > 5\text{min}) = \frac{3}{4} e^{-20\left(1-\frac{3}{4}\right)5/60}$$

$$=\frac{3}{4}e^{-\frac{25}{60}}=0.4944$$

$$P(T_q>10\,mins)=\frac{3}{4}e^{-\frac{50}{60}}=0.3259$$

e. Expected number of landings in a 20 minute period = 15/60×20 = 5.

3. Let us consider trains arrive at the yard every 15 minutes and the services time is 33 minutes. If the line capacity of the yard is limited to 5 trains, find the probability that yard is empty and the average no of trains in the system.

Solution:

λ = 1/15 per minute; μ = 1/33 per minutes

$$\rho=\lambda/\mu=2.2$$

Probability that the yard is empty:

$$P_o=\frac{\rho-1}{\rho^{N+1}-1}=\frac{2.2-1}{(2.2)^6-1}=1.068\%=0.01068$$

Average no of trains in the system:

$$L_s=\sum_{n-o}^{w}nP_n=P_1+2P_2+3P_3+4P_4+5P_5$$

$$=P_o\left[\rho+2\rho^2+3\rho^3+4\rho^4+5\rho^5\right]$$

$$=4.22$$

4. On average a single-processor CPU completes 10 jobs per second and a job arrives every 0.4 seconds. Jobs are executed immediately if the CPU is idle. Otherwise they will be placed in the hard disk to wait for their turns according to the order of arrivals. If arrival and completion times of jobs are exponentially distributed:

a) Estimate the average CPU time for each job.

1/10 = 0.1 second

b) Estimate the average total wait time for each job.

Solution:

$$\lambda=1/0.4=2.5\ /\ \text{seconds},\ \mu=10/\text{seconds},\ W_Q=\lambda/\mu\ (\mu-\lambda)=0.033\ \text{seconds}$$

M/M/1 with finite capacity (N): The condition $\lambda/\mu < 1$ is unnecessary in this case since the size of the queue is limited to N.

$$L=\frac{\lambda\left[1+N\left(\frac{\lambda}{\mu}\right)^{N+1}-(N+1)\left(\frac{\lambda}{\mu}\right)^{N}\right]}{(\mu-\lambda)\left[1-\left(\frac{\lambda}{\mu}\right)^{N+1}\right]}$$

$$W=\frac{L}{\lambda_a}$$

For arrival : $\lambda_a=\lambda$ For entering : $\lambda_a=\lambda(1-P_N)$

5.2.1 M/M/S & their Related Simple Problems

- Customers arrive according to a Poisson process with rate λ.
- The service times of customers are exponentially distributed with parameter μ.
- There are s servers, serving customers in order of arrival.

Stability Condition

$\lambda < s\cdot\mu$ or alternatively written:

The process $\{X(t), t \geq 0\}$, the number of customers in the system at time t, is again a continuous-time Markov chain with infinite state space.

The 'cut equations' are given by:

- $\lambda p_i - 1 = i\mu p_i,\ \ i=1,\ldots,s,$
- $\lambda p_i - 1 = s\mu p_i,\ \ i=s+1,\ldots.$

Hence,

$$p_i=\left(\frac{\lambda}{\mu}\right)^{i}\frac{1}{i!}p_0,\quad i=0,\ldots,s,$$

$$p_{s+k}=\left(\frac{\lambda}{s\mu}\right)^{k},p_s=\left(\frac{\lambda}{s\mu}\right)^{k}\left(\frac{\lambda}{\mu}\right)^{s}\frac{1}{s!}p_0,\quad k=0,\ldots$$

Finally, from the normalization equation $\sum_{i=0}^{\infty}p_i=1$ one can determine the unknown p_0.

Again, from the limiting distribution several long-run performance measures can be calculated.

Performance Measures in the M/M/s Queue

$\Pi_w =$ Probability that a customer has to wait:

$$=\sum_{k=o}^{\infty} p_{s+k} = \sum_{k=o}^{\infty} \left(\frac{\lambda}{s\mu}\right)^k p_s = \frac{p_s}{1-\rho}$$

B = expected number of busy server:

$$=\sum_{i=1}^{\infty} \min(i,s) p_i = \sum_{i=1}^{\infty} \left(\frac{\lambda}{\mu}\right) p_{i-1} = \frac{\lambda}{\mu}$$

$L_q =$ Expected number of waiting customer:

$$=\sum_{k=o}^{\infty} k p_{s+k} = \sum_{k=o}^{\infty} k \left(\frac{\lambda}{s\mu}\right)^k p_s = p_s \frac{\rho}{(1-\rho)^2}$$

$$W_q = L_q / \lambda, \qquad L = L_q + B, \qquad W = L/\lambda = W_q + 1/\mu$$

Problems

1. An airport has a single runway. Airplanes have been found to arrive at the rate of 15 per hour. It is estimated that each landing takes 3 minutes. How would the performance measures change if there are two runways while assuming the same arrival and service rates?

Solution:

a. Runway utilization:

Arrival rate = 15/hr (λ)

Service rate = 20/hr (μ)

No. of servers = 2 (s)

Utilization of each runway $= \rho = \lambda / s\mu = 3/8$

b. Expected number of airplanes waiting to land:

$$L_q = \frac{\rho \alpha^s p_o}{s!(1-\rho)^2}$$

$$\text{Note}: \alpha = s/\rho = 3/4$$

$$p_o = \left[\sum_{r=0}^{1} \frac{\alpha^r}{r!} + \frac{\alpha^s}{s!(1-\rho)}\right]^{-1}$$

$$= \left[1 + \frac{3}{4} + \frac{(3/4)^2}{2}\left(1 - \frac{3}{8}\right) - 1\right]^{-1}$$

$$= 0.4545$$

$$L_q = \left[\left(\frac{3}{8}\right)\left(\frac{3}{4}\right)^2 (0.4545)\right] / 2\left(\frac{5}{8}\right)^2$$

$$= 0.1227$$

c. Expected waiting time:

$$W_q = \frac{\alpha^s po}{s! s\mu (1-\rho)^2}$$

$$= \left[\left(\frac{3}{4}\right)^2 (0.4545)\right] / 2 \times 2 \times 20\left(1 - \frac{3}{8}\right)^2$$

$$= 0.00818 \text{ hrs} = 0.49 \text{ mins}$$

d. Probability that the waiting will be more than 5 mins? 10 mins? No waiting?

$$P(\text{no waiting}) = F_q(0) = 1 - \frac{\alpha^s p_o}{s!(1-\rho)}$$

$$= 1 - \frac{\left(\frac{3}{4}\right)^2 (0.4545)}{2(1 - 3/8)}$$

$$= 0.7955$$

$$P(T_q > t) = \frac{\alpha^s p_o}{s!\,(1-\rho)} e^{-s\mu(1-\rho)t}$$

$$P(T_q > 5\,\text{min}) = \frac{(3/4)^2(0.4545)}{2(5/8)} e^{-2\left(\frac{1}{3}\right)\left(\frac{5}{8}\right)5}$$

$$= 0.1245$$

$$P(T_q > 10\,\text{min}) = 0.0155$$

e. Expected number of landings in a 20 minute period = 15/60 × 20 = 0.5

[The departure process is Poisson with parameter λ]

2. A bank has established two counters one for commercial banking and the second for personal banking. Arrival and service rates at the commercial counter are 6 and 12 per hours respectively. The corresponding numbers at the personal banking counter are 12 and 24 respectively. Assume that arrivals occur in Poisson processes and service times have exponential distributions.

(i) Assuming that the two counters operate independently of each other determine the expected number of waiting customers and their mean waiting time at each counter.

	Commercial	Personal
λ	6/hr	12/hr
μ	12/hr	24/hr
$\rho = \frac{\lambda}{\mu}$	0.5	0.5
$L_q = \frac{\rho^2}{1-\rho}$	0.5	0.5
$W_q = \frac{\rho}{\mu(1-\rho)}$	5 mins	2.5 mins.

(ii) What is the effect of operating the two queues as a two server queue with arrival rate 18/hr and service rate 18/hr? What conclusions can we drawn from this operation?

Solution:

	Two – server queue
λ	18/hr
μ	18/hr
Number of servers (s)	2
$\rho = \frac{\lambda}{s\mu}$	·5

$\alpha = \frac{\lambda}{\mu}$	1
$P_o = \left[\sum_o^1 \frac{\alpha^r}{r!} + \frac{\alpha^2}{2(1-\rho)}\right]^{-1}$	0.4
$L_q = \frac{\rho\alpha^2 P_o}{2(1-\rho)^2}$	0.4
$W_q = \frac{\alpha^2 P_o}{(2)2\mu(1-\rho)^2}$	1.33 mins

The two-server queue operation is more efficient than the two single-server operations.

Incidentally, the efficiency of multi-server queues over single server systems is the reason that multi-server service systems, whenever possible, use single waiting lines feeding multiple counters for service; e.g. airline checking counters, checkout counters in stores effectively operate this way because of jockeying among the waiting lines.

Permissions

All chapters in this book are published with permission under the Creative Commons Attribution Share Alike License or equivalent. Every chapter published in this book has been scrutinized by our experts. Their significance has been extensively debated. The topics covered herein carry significant information for a comprehensive understanding. They may even be implemented as practical applications or may be referred to as a beginning point for further studies.

We would like to thank the editorial team for lending their expertise to make the book truly unique. They have played a crucial role in the development of this book. Without their invaluable contributions this book wouldn't have been possible. They have made vital efforts to compile up to date information on the varied aspects of this subject to make this book a valuable addition to the collection of many professionals and students.

This book was conceptualized with the vision of imparting up-to-date and integrated information in this field. To ensure the same, a matchless editorial board was set up. Every individual on the board went through rigorous rounds of assessment to prove their worth. After which they invested a large part of their time researching and compiling the most relevant data for our readers.

The editorial board has been involved in producing this book since its inception. They have spent rigorous hours researching and exploring the diverse topics which have resulted in the successful publishing of this book. They have passed on their knowledge of decades through this book. To expedite this challenging task, the publisher supported the team at every step. A small team of assistant editors was also appointed to further simplify the editing procedure and attain best results for the readers.

Apart from the editorial board, the designing team has also invested a significant amount of their time in understanding the subject and creating the most relevant covers. They scrutinized every image to scout for the most suitable representation of the subject and create an appropriate cover for the book.

The publishing team has been an ardent support to the editorial, designing and production team. Their endless efforts to recruit the best for this project, has resulted in the accomplishment of this book. They are a veteran in the field of academics and their pool of knowledge is as vast as their experience in printing. Their expertise and guidance has proved useful at every step. Their uncompromising quality standards have made this book an exceptional effort. Their encouragement from time to time has been an inspiration for everyone.

The publisher and the editorial board hope that this book will prove to be a valuable piece of knowledge for students, practitioners and scholars across the globe.

Index

www.ingramcontent.com/pod-product-compliance
Lightning Source LLC
LaVergne TN
LVHW080601210926
848485LV00003B/89
* 9 7 8 1 6 3 9 8 9 7 1 5 5 *